Flexible Printed Circuitry

Thomas H. Stearns
Brander International Consultants
High Technology Services
Nashua, New Hampshire

McGraw-Hill

New York San Francisco Washington, D.C. Auckland Bogotá
Caracas Lisbon London Madrid Mexico City Milan
Montreal New Delhi San Juan Singapore
Sydney Tokyo Toronto

Library of Congress Cataloging-in-Publication Data

Stearns, Thomas H.
　　Flexible printed circuitry / Thomas H. Stearns.
　　　　p. cm.—(Electronic packaging and interconnection series)
　　Includes index.
　　ISBN 0-07-061032-0 (hc)
　　1. Flexible printed circuits. I. Title. II. Series.
TK786.P7S74 1995
621.3815′31—dc20　　　　　　　　　　　　　　　　95-38782
　　　　　　　　　　　　　　　　　　　　　　　　　　　　CIP

McGraw-Hill

A Division of The McGraw·Hill Companies

Copyright © 1996 by The McGraw-Hill Companies, Inc. All rights reserved. Printed in the United States of America. Except as permitted under the United States Copyright Act of 1976, no part of this publication may be reproduced or distributed in any form or by any means, or stored in a data base or retrieval system, without the prior written permission of the publisher.

1 2 3 4 5 6 7 8 9 0　DOC/DOC　9 0 0 9 8 7 6 5

ISBN 0-07-061032-0

The sponsoring editor for this book was Stephen S. Chapman, the editing supervisor was Paul R. Sobel, and the production supervisor was Suzanne W. B. Rapcavage. It was set in Century Schoolbook by Victoria Khavkina of McGraw-Hill's Professional Book Group composition unit.

Printed and bound by R. R. Donnelley & Sons Company.

Information contained in this work has been obtained by The McGraw-Hill Companies, Inc. ("McGraw-Hill") from sources believed to be reliable. However, neither McGraw-Hill nor its authors guarantees the accuracy or completeness of any information published herein and neither McGraw-Hill nor its authors shall be responsible for any errors, omissions, or damages arising out of use of this information. This work is published with the understanding that McGraw-Hill and its authors are supplying information, but are not attempting to render engineering or other professional services. If such services are required, the assistance of an appropriate professional should be sought.

McGraw-Hill books are available at special quantity discounts to use as premiums and sales promotions, or for use in corporate training programs. For more information, please write to the Director of Special Sales, McGraw-Hill, 11 West 19th Street, New York, NY 10011. Or contact your local bookstore.

 This book is printed on recycled, acid-free paper containing a minimum of 50% recycled de-inked fiber.

This book is dedicated to Sid, Vic, Ben and all those who labored, through the years, to create and establish flexible printed wiring as the universal interconnect technique; to the guys and gals who never stopped looking for a better way and never could answer the question, "what kind of engineer are you?"

Grateful appreciation is extended to Marc Pare and Signart Co. for sketches and line drawings, and to the companies that opened their doors to the author for equipment and facilities photographs.

Contents

Preface   xi

Chapter 1. Introduction — 1

Purpose — 1
Introduction — 1
History — 2
Status — 3
Descriptions — 4
Summary — 8

Chapter 2. The Engineered Connection — 9

Comparison with PWBs and Wire — 9
Cost — 10
Weight and Space Savings — 12
Dynamics, Installation — 13
Environmental — 14
Troubleshooting, Repair — 16
Summary — 17

Chapter 3. Manufacture of Flexible Printed Wiring and Printed Wiring Boards — 19

Introduction — 19
Comparison of PWB and FWB Manufacturing — 19
Transparency Stability — 20
Handling — 24
Coverlayers — 28
Summary — 28

Chapter 4. Design — 29

Introduction — 29
Data Requirements — 30
Intellectual Approach — 30

Procedure	31
Refinements	38
Tooling	45
Rigid Flex	48
Materials	51
Composites	66
Shielding	70
Controlled Impedance	73
Summary	83
Chapter 5. Dielectric Materials	**85**
Introduction	85
Elements	85
Adhesives	87
Films	93
Comparison Charts	98
Potting	100
Summary	102
Chapter 6. Conductive Materials	**103**
Copper Foils	103
Other Conductive Materials	109
PTF and Conductive Inks	110
Shielding Materials	111
Summary	112
Chapter 7. Adhesiveless Materials	**113**
Introduction	113
Definition	114
Construction Materials	114
Rigid Flex	117
Benefits	117
Comparisons	118
Summary	124
Chapter 8. Manufacturing Processes	**125**
Introduction	125
Manufacturing Overview	125
Materials	127
Surface Finish	137
Lamination	138
Imaging	143
Etching	152
Stripping	159
Outline	160

Tooling	160
Process Sequence	165
Summary	183

Chapter 9. Rigid-Flex Manufacture — 185

Introduction	185
Definitions	186
Process	190
Materials	204
Photographs	213
Summary	216

Chapter 10. Standards and Specifications — 217

Introduction	217
Cost Impact of Quality Requirements	218
Sources	218
Double Dimensioning	219
Document Systems	220
Guidelines	221
Summary	223

Chapter 11. Assembly — 225

Background	225
Joining Processes	226
Forming	236
Potting	238
Conformal Coat	240
Nomenclature	241
Mechanical	242
Summary	243

Chapter 12. Examples of High-Volume and Unusual Flexible Printed Wiring — 245

Military Reel Cable	245
High-Volume Epoxy-Mat FPW	246
High-Volume Vinyl FPW	247
Stripline	247
High-Voltage Ignitor Cable	249
Vacuumtight Stripline	249
Outsized Backplane	250
High-Volume Miniature Backplanes	254

Chapter 13. Summary — 257

Purpose	257
Chapter 1: Introduction	257

Chapter 2: The Engineered Connection — 258
Chapter 3: Manufacture of Flexible Printed Wiring and Printed Wiring Boards — 259
Chapter 4: Design — 260
Chapter 5: Dielectric Materials — 262
Chapter 6: Conductive Materials — 263
Chapter 7: Adhesiveless Materials — 264
Chapter 8: Manufacturing Processes — 265
Chapter 9: Rigid-flex Manufacture — 269
Chapter 10: Standards and Specifications — 271
Chapter 11: Assembly — 272
Chapter 12: Examples of Unusual and High-Volume Flexible Printed Wiring — 273

Appendix A. Foreign Material — 275

Appendix B. Adhesiveless Materials in Dewar Circuitry — 277

Glossary   279
Index   287

Electronic Packaging and Interconnection Series
Charles M. Harper, Series Advisor

ALVINO • *Plastics for Electronics*
CLASSON • *Surface Mount Technology for Concurrent Engineering and Manufacturing*
GINSBERG AND SCHNORR • *Multichip Modules and Related Technologies*
HARPER • *Electronic Packaging and Interconnection Handbook*
HARPER AND MILLER • *Electronic Packaging, Microelectronics, and Interconnection Dictionary*
HARPER AND SAMPSON • *Electronic Materials and Processes Handbook*, 2/e
LICARI • *Multichip Module Design, Fabrication, and Testing*
SERGENT AND HARPER • *Hybrid Microelectronics Handbook*, 2/e
SOLBERG • *Surface Mount Guidelines*, 2/e

Related Books of Interest

BOSWELL • *Subcontracting Electronics*
BOSWELL AND WICKAM • *Surface Mount Guidelines for Process Control, Quality, and Reliability*
BYERS • *Printed Circuit Board Design with Microcomputers*
CAPILLO • *Surface Mount Technology*
CHEN • *Computer Engineering Handbook*
COOMBS • *Printed Circuits Handbook*, 4/e
DI GIACOMO • *Digital Bus Handbook*
DI GIACOMO • *VLSI Handbook*
FINK AND CHRISTIANSEN • *Electronics Engineers' Handbook*
GINSBERG • *Printed Circuits Design*
JURAN AND GRYNA • *Juran's Quality Control Handbook*
MANKO • *Solders and Soldering*, 3/e
RAO • *Multilevel Interconnect Technology*
SZE • *VLSI Technology*
VAN ZANT • *Microchip Fabrication*

To order or receive additional information on these or any other McGraw-Hill titles, in the United States please call 1-800-822-8158. In other countries, contact your local McGraw-Hill representative. **BC14BCZ**

Flexible Printed Circuitry

Preface

This is a book about flexible printed circuitry which follows its development from an accidental origin in a materials experiment to present status as an important aid to low-cost electronic equipment manufacture. Until recently only a minor factor in engineered interconnections, regarded by most packaging engineers as a curious but limited outgrowth of printed wiring board technology, flexible printed wiring is at last enjoying the attention and utilization which it deserves. This highly engineered, mass-produced product brings low assembly cost and repeatable, consistent performance to any electronic device. Its use guarantees high first-pass inspection yield, minimum rework, simplified field service, least weight and highest circuit density.

Development of production technology is traced from initial production in fusion-bond thermoplastic dielectrics to wet adhesives and high temperature dielectric films; thence to coatings of B-stage magnet-wire varnish and proprietary blends of thermoplastic and thermosets and into currently popular adhesiveless composites of conductor and dielectric. Unique terms are defined, constructions are explained and interactions between design, materials and manufacturing technology are presented with the intention of providing a comprehensive reference book to guide program managers, purchasing and engineering interests. Two chapters treat the subject of special issues of materials, design and manufacture associated with rigid-flex circuitry.

Flexible printed wiring is an ever-changing, exciting technology which plays strongly into current trends in electronic packaging. The materials and manufacturing methods have enormous development potential for use in MCM's, PCMCIA cards and other huge new product areas where extreme thinness, highest density and volume manufacturability are important.

The trend to increasing use of flexible circuitry is powered by relentless demand for more circuit functionality coupled with a steady drop in component size and power demand. Taken together, these factors marginalize the significance of mechanical strength—the primary virtue of traditional "rigid" printed wiring boards—and emphasize the importance of unlimited adaptability—the strong point of flexible printed wiring. As mechanical strength recedes in significance, packaging engineers realize that flexible circuitry can do it all: it's flexible not only in the sense that it is easily folded and bent to rout throughout three-dimensional packages, but because it can both support and interconnect semiconductor devices, seamlessly interconnect dynamical devices like printed heads, flying disc readers and test probes, and simultaneously accommodate unavoidable variations in position and dimension which crop up in complex assemblies.

There are penalties associated with use of flexible circuitry, of course. Engineering costs and cycle time delays are painfully large; once the design is complete, changes and corrections require careful documentation and add more cost. Any system which reduces random error and promotes repeatability is complicated to set up and difficult to change. But when manufacturing volume is significant and circuit density is high, flexible printed wiring is the clear choice for assembly automation. This book explains the important factors and their interactions, and gives the reader a solid overview of the usefulness and promise of this technology.

Thomas H. Stearns

Chapter 1

Introduction

Purpose

This is a book about flexible printed wiring. Its purpose is to explain flexible printed wiring—materials, manufacturing methods, design—to packaging engineers and procurement specialists so that they can more effectively exploit this powerful interconnect technique. In the following pages, program managers and purchasing professionals will gain insight into the interrelationship of design, manufacture, and cost of procurement.

Although rigid and flexible printed wiring products appear to be much alike and do share many technologies and manufacturing equipments, the fundamental differences, rooted in the materials of construction, give rise to enormous distinctions. We hope to make it all clear to you in the following chapters.

Introduction

It's been almost 40 years since flexible printed wiring first came over the horizon of the electronics scene. Like many other high-technology products, this one was first nurtured and exploited by the military, where technical advantage outweighs cost. Commercial interests, if belatedly and begrudgingly, eventually recognized the value of this interconnector; today, sales volume of flexible printed wiring is about 88% commercial, 22% military.

With the late-'80s' downturn in defense expenditures, the steady upward sales trend flattened; nevertheless, recent surveys show almost $500 million in yearly United States sales and $1.8 billion worldwide. Flexible printed wiring is, at least commercially, mature.

History

The flexible printed wiring product came into commercial use at Sanders Associates, Inc., Nashua, New Hampshire, in the mid-1950s, growing out of an attempt to find out why printed wiring boards (PWBs) cracked when they were dipped in solder. During an investigation of the problem it was noted that laminates were produced by bonding copper foil to a paper-phenolic substrate with a film of butyral-phenolic adhesive. Thinking he could avoid cracks by eliminating the brittle substrate, Victor Dahlgren, a Sanders components engineer, bonded copper foil directly to butyral-phenolic film to produce a laminate and made etched "circuits" from it. Butyral phenolic is an adhesive which, by itself, has little structural strength. Thus Vic's experimental circuits, although crackless, were flimsy, almost uselessly delicate.

Royden Sanders, president of the company, saw these superthin printed circuits not as just a limp first attempt at avoiding cracks in rigid boards, but as a practical, manufacturable product which could replace wire harnesses. He urged Vic to carry on, to "go find a better film"—something tough and easy to handle—and to use a second ply of film to insulate over the conductors. This "covercoating" idea also came from another project in which a plastic film was bonded onto the top surface of a printed wiring board to act as a solder mask and insulator.

Other scenarios for the creation of flexible printed wiring have been suggested by historians. Whoever is correct, the critical moment came with the realization that printed circuitry made from flexible dielectrics can replace wiring harnesses and that short circuits in a severely deformed flexible printed circuit can be avoided by covercoating or coverlayering with a second dielectric film.

By 1955 the outlines of this new interconnect industry had been drawn. The first commercial process was to fuse Kel-F* (later FEP Teflon®,† vinyls, polyurethanes) directly onto an oxide-treated copper foil, by application of sufficient pressure and temperature, to make a base laminate. The laminate was processed through typical printed-circuit image and etch steps to form a *base circuit*. Like Vic's experimental PWBs, the new product was then encapsulated or "coverlayered" with a second "autogenously bonded" (fusion-laminated) layer of film.

Adhesion of film to conductor was enhanced by use of oxide treatment, which both passivated the copper surface to extend useful life at elevated temperatures and created a deep microstructure for good mechanical interlock between conductor and insulation. Access to etched conductors at terminal areas was provided by windowing, or cutting away those areas of the coverlayering film; these weakened areas were reinforced by

---

*Tradename for a fluorocarbon film produced by W. L. Shamban Co.
†Registered trademark of Dupont.

application of "backers" of film built into the base area. Later, as the technology developed, smaller, drilled openings which individually exposed each terminal were used in place of the windows.

The next few years saw development of every aspect of the technology from design through roll-to-roll manufacturing. Each application was worked from both ends—cooperatively, with customer engineers to develop a design that fitted the package, met electrical requirements, and facilitated efficient manufacturing and test, then within the Sanders production group to develop equipment and techniques to build it. This was innovation in the most efficient possible fashion: concurrent development of both need and means of production to a degree that probably won't be seen again.

Things moved quickly; by 1959 a feature article in the *Boston Globe* identified the new product—now called *Flexprint*—as "perhaps the hottest new product" of Sanders Associates. The incredible level of activity and creativity resulted in more than 40 Flexprint patents by 1965; of the first 100 patents granted to Sanders Associates, almost half were products of the Flexprint group. Included were U.S. 3,201,851, "provision of drilling and electrically conductive surface treatments . . . in multilayer circuits" (PTH process), and U.S. 3,383,564, "three-dimensional printed circuit matrices having one or more flexible circuit flaps coextensive with the interconnected layers" (rigid-flex circuitry).

Status

Yet in one important respect the technology of flexible printed wiring has never matured. Rushed too fast from laboratory to commercialization, the product is still manufactured by techniques which are nothing more than scaleups of the first experiments; compared to rigid-substrate printed wiring, flexible printed wiring (FPW) is far less cost-predictable and process-mature. This is proven by the fact that quotations for printed wiring boards (PWB) are available within days of request, and from multiple vendors will fall in a small cluster of price and delivery. Indeed, some companies even advertise by publishing a cost matrix linking size and purchase quantity which allows prospective customers to instantly calculate their own prices! By contrast, quote response for flexible product is much longer, prices will range more widely, and product delivery will be both longer as an estimate and less certain in reality, reflecting a manufacturing technology which is less well-developed.

A good part of the blame for this can be laid to the inherent nature of "flexible" versus "rigid" materials. Dimensional stability of rigid laminates is better, a vitally important advantage which allows larger panel sizes and tighter tolerances. Fusion bonding, used in the first few years of flexible circuit production, was a magnificent process for preserving electrical properties, but always resulted in poor dimensional stability.

The FPW industry recognized the manufacturing inefficiencies which the fusion process inflicts. Because the primary factor determining dimensional stability is lamination temperature, liquid, low-temperature curing adhesives replaced the fusion process as a way to join foil to polymer film in the early '60s; soon this technique gave way to machine-coated, dry-to-the-touch (B-stage) adhesive dielectric systems which processed at higher temperatures but improved uniformity and reduced lamination and cleanup labor. Adhesive-coated dielectrics continue to be the main workhorse in FPW production, but in spite of decades of development still lack adequate stability: until the recent advent of so-called adhesiveless composites, the FPW industry has never enjoyed stability in flexible dielectrics comparable to that of rigid materials. Laminate stability (and rigidity) produce higher yields and more predictable production because less product is lost to out-of-tolerance features or handling damage.

It's fair to say that major obstacles to the development of efficient FPW manufacturing technology are embedded in the fundamental properties of the materials. This is a product which hasn't reached full potential, not because of lack of effort, but just because of its nature. The problems to be overcome before flexible printed wiring becomes a commodity like rigid printed wiring are very challenging. The recent emergence of dimensionally stable, directly metallized flexible materials concurrent with an enormous rush to thin, fine-featured multilayer constructions will powerfully accelerate the exploitation of this technology. It's been coming for 40 years.

Descriptions

To best clarify the discussion in other chapters, it is appropriate that we set forth definitions, descriptions, and nomenclature which are peculiar to this product and its manufacturing technology. Words which are used in an FPW-unique sense are italicized in the following overview and are defined in the Glossary.

Definition of flexible printed wiring

The IPC* defines flexible printed circuitry as a "random arrangement of printed wiring utilizing a flexible base material with or without cover layer." This definition evokes the familiar printed circuit board

---

*Institute for Interconnecting and Packaging Electronic Circuits.

made on a flexible substrate or *laminate* (a composite of *dielectric* and conductor *foil*) and is immediately meaningful to those who are familiar with PWBs. (It should be understood that the term "flexible printed circuitry" is more accurately used to denote FPW assembled to its associated connectors and other components.)

For greater clarity the IPC definition requires expansion as follows:

- The conductor pattern is created by a repetitive manufacturing process such as printing a conductive ink, imaging and etching or mechanically cutting a metal foil, or forming and placing wire by numerical control (NC) equipment.
- The dielectric layers are bonded tightly to the conductor pattern.
- Overall thickness is typically less than 0.02 in.

This added terminology clears up the confusion left by *printed* and *random* and increases the range of manufacturing methodology.

Construction

A flexible printed circuit consists of three layers of material: a *base layer* of dielectric, a central *conductor layer* and a top dielectric layer called a *coverlayer* (if a film) or *covercoat* (if a liquid coating). The coverlayer or covercoat may be absent (in low-cost circuitry) or may be a photoimaged or screen-printed polymer coating.

Openings or *apertures* are provided in the coverlayer to allow contact with the conductor layer at desired *terminal* or *pad* locations. The process of producing apertures through a dielectric is called *baring*; making apertures through the base dielectric either before or after the conductor pattern is created is called *reverse baring*. If a pad is bared on both sides it is *unsupported* or *fully bared*.

Coverlayer films are aligned to the etched features and *tacked* in place to maintain alignment until the lamination process. Tacking can be accomplished by localized heat and pressure (modified soldering irons are commonly used) or by moistening the coverlayer *adhesive* with solvent, then pressing the layers together for a brief moment.

Conductor patterns are defined by photomechanical reproduction of a dimensionally accurate master photopattern called *artwork*. The method of creating the conductor pattern influences the materials used and the manufacturing methodology. Most FPW is constructed by *subtractive etching* or removing unwanted parts of a foil cladding to leave etched foil conductors on an adhesive-bonded dielectric base consisting of dielectric, adhesive, and foil. For roughly 78% of FPW, the dielectric is a high-performance, high-cost polyimide film from

0.001 to 0.002 in in thickness. Most of the remaining 22% utilizes polyester or other low-cost dielectrics.

A rising percentage of FPW is produced by *semiadditive processing*. The procedure is to image a thinly metallized *seed layer* of laminate, exposing the desired conductor pattern, then, using the seed layer as an interconnecting *busbar*, plate this pattern up to desired conductor thickness. Finally, the resist image is stripped off and the seed layer of metal is removed by a brief *differential etch* process.

Performance of high-quality dielectrics is degraded by the adhesive joining the dielectric and conductor layers; for more demanding use a newly emerging class of composites, called *adhesiveless,* is employed. These materials are produced by several processes:

- Casting a liquid dielectric precursor onto an existing foil and curing to create the dielectric layer.
- Depositing conductive metal onto an existing dielectric film.
- Utilizing a high-temperature, "high-performance" adhesive, whose properties approach those of the dielectric, to bond foil to dielectric.

All these methods eliminate highly plasticized adhesives and thus enhance performance (see Chap. 7).

Inexpensive circuitry can be created by printing a *conductive ink* directly onto a dielectric film to form the conductor pattern. This is the polymer thick film (*PTF*) method; these patterns may be insulated by a *screen-printed* dielectric layer which has openings at desired interconnection points; additional conductor layers can then be applied to create multilayer structures. Two major disadvantages to this technique are (1) high circuit resistivity (10 to 30 times higher than copper) and (2) necessity to use unusual termination techniques.

There is rapidly increasing interest in the use of conductive adhesives for mounting high-density semiconductor devices, because this process requires only modest heat exposure, does not require cleaning, and is adapted to *fine-pitch* (less than 0.02 in on centers) terminal patterns. Conductive adhesives are required for PTF patterns; another possible terminating technique suitable for these nonsolderable circuits is use of zero insertion force (*ZIF*) connectors or other very low contact force pressure means.

Terminations

It's typical for each *pad*—termination site, an enlarged area on a conductor, usually at an end—to have a *throughhole* to receive a connector pin or other hardware item which is soldered to the pad. Most throughholes holes are created either by NC drilling or die punching.

As circuit complexity increases and more than one conductor layer is required, plated-throughhole (*PTH*) technology is employed to simplify the assembly process. By means of complex chemical processes, the cylindrical walls of a throughhole in the FPW are made conductive, then plated with copper to form a "barrel" which electrically interconnects all the pads, in whatever layer, that are pierced by the hole. Attachment between the barrel and the cylindrical drilled edge of an inner pad is called an "interfacial bond." Barrels terminate at pads on the two outer surfaces; attachment to either pad interconnects to all pads at that hole location.

The most common use for PTH is to interface a connector to the FPW; connector pins are inserted through a cluster of PTH barrels and soldered. The mechanics of this interface are favorable to reliability because the pins and PTH barrels form the classic plug-in-hole construction which protects the joints from stress.

A major production issue in all FPW is *registration* or *concentricity* of hole to terminal to *covercoat opening,* the so-called *annular ring* problem (see Fig. 4-7). Concentricity and registration are controlled by tooling methods which include:

1. Preimage *tooling holes*
2. *Postetch* (*PE*) optical alignment of tooling hole to etched pattern
3. Local realignment or *cluster registration*
4. *Eyeball* or operator local *alignment*

Whether in *panel* or continuous roll form, FPW is normally produced in a tightly nested pattern called a *composite,* which has as many *repeats* of the *circuit pattern* as will fit onto the panel or *roll-to-roll* image area.

As much of the process as possible is performed in panel form, because this reduces handling labor. Once through the coverlayer lamination cycle, individual circuits are cut from the panel or *outlined.*

When the FPW customer intends to use *panelized* assembly, circuits may be partially outlined (leaving tiny *tugs* holding the circuits in place, to be torn out of the panel after assembly) or fully outlined but *returned to panel* and shipped in panel form.

The most common outlining method is by *steel rule die* (a low-cost cutting tool consisting of knife-edge steel blades inserted into a backer) cutting; other methods include routing by NC equipment or, for highest precision, matched die cutting. Outline tools are *aligned* to the circuits by secondary tooling holes, *fiduciaries,* templates, or eyeball.

Surface finish

Contact with FPW pads is usually effected by soldering; low insertion force (LIF) and ZIF connectors are also used. All require special preparation of the pad surface; because it's easier to do this in panel form, and because of shelf-life concerns, the FPW manufacturer usually provides this service, called *surface finish*.

A common treatment is hot-air leveled or plated-and-reflowed tin-lead solder; these coatings preserve solderability for many months and are compatible with low-cost connectors. Organic antitarnish or antioxidant coatings [organic protective coatings (OPC)] are also used, and have the considerable advantage that they do not require high temperatures.

Nomenclature

It's a common practice in the printed circuit industry to build manufacturer's names or logos, part numbers, date codes, and other text information or *nomenclature* into the artwork which is used to define the conductor pattern. Other nomenclature items which appear on FPW are connector pin and test-point designations and design revision level. The advantage of building this information into the artwork is that it is then produced as an integral part of the circuitry to assure accuracy and clear, permanent identification. Nomenclature may also be applied to finished circuitry by silk-screen printing or rubber stamping, usually with an epoxy or other durable ink.

Summary

Flexible printed wiring is a development from rigid printed wiring technology. Commercialized in the mid-50s, it has been steadily developed and improved until today, global sales exceed $1.3 billion yearly.

The first production circuits were manufactured by fusing high-performance polymer films onto copper foil. Dimensional instability in the fusion process led to the development of binary dielectrics including adhesive-bonded film systems.

Many special terms, and common terms with unusual meanings, are found in FPW technology. A list of these terms with brief explanations of their meaning is found in the Glossary.

Chapter

2

The Engineered Connection

Comparison with PWBs and Wire

The purposes of flexible printed wiring—henceforth FPW—are to lower cost and enhance performance. Of the three most common products used for *macro level* interconnections—wire harnesses, FPW, and printed wiring boards (PWBs)—the most universal, by a wide margin, is FPW.

Both FPW and PWBs are highly engineered, volume-produced interconnects which share the advantages of low assembly cost and consistent transfer functionality. PWBs are better where component support is the important function; FPW is usable in these applications provided it is locally stiffened and the components aren't too heavy. In all other applications, FPW is the clear choice over PWBs because it absorbs dimensional variation, accommodates the most extreme three-dimensional layout and installation protocols, and absorbs shock and vibration within its compliant structure. In most applications FPW is also superior to wire harnessing; for applications such as hard disk drives which have severe flexural life and torque specifications, FPW is the only choice. Given the trend toward ever smaller components, down to chip-on-board, and the rapid penetration of FPW technology into the thin, multilayer microinterconnect domain (MCM-L), FPW can be predicted, as Roy Sanders foresaw almost 40 years ago, to become the dominant interconnect technique.

We can broadly illustrate the advantages of FPW by an example in which the purpose is to interconnect a PWB with a panel; in this instance FPW provides a right-angle interconnect path and accommodates locational variations between PWB and panel. This function could also be performed by a wire harness. The advantages of FPW are:

- It is available from inventory at a fixed price, each piece (within tolerances) an exact copy of all others.
- Because of designed-in keying features, the termination areas of the FPW assembles to the PWB and panel hardware in an unmistakable way, reducing assembly labor and the likelihood of wiring error and rework labor.
- Conductor placement is identical in every assembly.
- Planar layout and consistent terminal location facilitate electrical testing.
- Consistent shape fits the assembly accurately, providing sufficient service loopage for assembly and field disassembly without concentrated stress on any run. In comparison, when a wire harness is installed and the conductors are mechanically secured to the package structure, there is always danger of vibration-induced failure because the shortest strand bears the greatest load.
- Use of flat conductor or other parallel conductor harnessing reduces the risk of stress concentration but also reduces ease of installation, since custom tailoring to the mechanical or wire-routing singularities of the application is not possible.
- FPW produces better looking, more organized, more esthetically pleasing assemblies.

When low-level signals are involved, FPW provides an important added cost saving over wire harnesses, in which variable conductor length and uncontrolled location of a disturbing wire to a sensitive wire forces use of excessive shielding to protect against worst-case proximity. With FPW interconnection, if preproduction testing shows acceptable performance without shields, the consistency of performance in production FPW assures that the same simple, least-cost design can used without fear of excessive crosstalk. Even when testing shows that shielding is required, it is easily added, reliable, and cost-effective.

Cost

FPW has a purchase price (what you pay to get it into inventory) and an assembled cost (the sum of price plus costs of installation and test). Wire harnesses, relatively inexpensive to buy, are built up from individual conductors and consume labor according to the number of ends to be stripped, identified, and terminated. FPW is a manufactured assembly of conductors, a single entity which is mass-terminable and self-identifying because position equates to identity—the same conductor run is always in the same position.

Justifying the use of FPW through reduction in assembly cost is a difficult task, not because the benefit is small, but because the information is hard to get. The FPW purchase price is highly visible at the procurement level but establishing accurate assembly, test, and rework costs for objective comparison of FPW to wire harnesses is difficult. It takes work, but when all factors are added up, FPW lowers the cost of electronic assemblies because it's easier to install.

FPW disadvantages

FPW is not a catalog item. Each application is unique; FPW for a given application must be designed, purchased, expedited, and inventoried as a separate line item. Because each FPW design requires considerable up-front engineering effort and heavy investment in tooling, both "hard" and data-based, the product is difficult to change and costly to procure in small quantities. Further, because procurement can begin only after the interconnect design is completed, significant program delays are unavoidable. In contrast, wire harnesses can be built with virtually no tooling, can be started before a run list is fully developed, and can be reworked to add conductors and change run-to-pin connections at any time.

Where wire is best

FPW is the technique of choice for assembly in volume; wire is best for prototypes and experimental use. The cost of FPW is primarily driven by circuit area; that of wire harnesses by the number (and length) of conductors. Where there are many conductor runs in a small area, FPW is the winner because cost per run is low. If there are few conductors and long runs, wire is the choice. The table below compares the two for a variety of criteria.

	FPW	Wire
Engineering start-up cost	High	Low
Response time	Long	Short
Purchase cost	High	Low
Assembly labor	Low	High
Inspection cost	Low	High
Rework	Low	High
Consistency	High	Low
Shield requirements	Least	Most
Ease of change	Low	High
Cost basis	Area	Number of conductors

If a packaging engineer or designer is interested in volume assembly cost, FPW is a strong contender. Provided that the expected production run justifies front-end engineering and tooling costs, FPW will lower assembled cost.

The level of reduction in cost rises with increasing utilization and design accommodation to FPW. That is, least cost results when:

- FPW is used for all interconnect levels: component to component, component group to group, and panel to group
- All hardware such as connectors and larger components is FPW-adapted
- The FPW design allows mass termination [wave solder, IDCs (insulation displacing connectors)]
- The FPW design anticipates (provides for) final electrical test.

Inspection, rework

Test/inspection records show that labor-intensive wire harnesses account for a high percentage of rework activity. That's because, in any manufacturing activity, each degree of freedom is always accompanied by risk of error. Wire harnesses are built up by bundling together a group of individual conductors; examples of resulting harness variables (and potential sources of error) are:

- Number of conductors
- Conductor length
- Size of conductor
- Conductor placement
- Interconnect pattern
- Stripping damage

With the exception of conductor size, which varies when image and etching processes aren't well-controlled, FPW eliminates all of these variables and therefore minimizes rework.

Weight and Space Savings

FPW has enormous value in its ability to reduce volume, weight, and dynamic loading and to compress complex circuitry into buildable, serviceable packages. Much of FPW's volumetric efficiency springs from careful engineering of conductor size and dielectric materials. To explain:

- Wire sizes are commonly dictated by handling issues arising from termination rather than electrical demands such as resistivity or current-carrying ability. This is particularly true inside packages—the principal site for FPW use—where current loadings are small and resistivity is not critical. In both interconnect techniques the greatest stress concentration occurs at terminations. Harness wire must be everywhere the same size; a choice which is oversize to allow for safe termination results in oversize midbody conductors. FPW has engineered, stress-distributing terminations: big where attachment is made, then tapering to a midbody which is scaled—much smaller—to carry the electrical load. Since FPW is a structurally unitized product in which the dielectric system and each conductor run reinforce and support all runs, it is safe to reduce low-current conductors to sizes which are dictated by the ability to image and etch; this accounts for a major part of the increased efficiency of weight and space utilization.

- When current-carrying capacity is a factor, the rectangular FPW conductor shape has important advantages. Current-carrying capacity is limited by heat buildup. Compared with round wire of equivalent cross section, rectangular FPW conductors have more surface area for improved heat rejection. This factor, plus reduced insulation thickness, allows FPW to carry greater current, size for size, than round wire.

- Very often there is sufficient room under, between, and above components to route FPW without space allowance because its rectangular shape inherently packs more efficiently than round. Where it is necessary to provide a dedicated channel, the rectangular form factor fits more gracefully with less design disruption than the round cross section required for wire harnessing.

- Reduced volume means reduced weight, given similar materials (both interconnects use copper conductors and polymeric insulation). FPW occupies substantially reduced volume, compared with wire harnessing, and correspondingly has reduced weight.

- In air-cooled packages, FPW can be designed and installed for least obstruction of flow.

Dynamics, Installation

Wire bends with equal ease in all directions. FPW is relatively more flexible in one plane and relatively stiffer in others. Wire can be reformed at will. FPW has a manufactured shape, in its major plane, which is very unformable; it must be correctly designed and sized.

When it is engineered to take advantage of the reduced-stiffness direction, FPW is easier to install.

FPW is uniquely suited to dynamic (moving) use provided the design exploits its enhanced flexibility. Disk-drive systems could not be built economically without the long-life FPW which connects with the flying head. Another example is the interconnecting circuit in a dot-matrix printer: one end of the FPW terminates on the chassis, the other end, following a U pathway, terminates at the traveling head. Motion of the printer causes the FPW to roll on itself as the head moves so that the stress is distributed by the naturally formed bending radius and is absorbed by the conductor and dielectric.

Another good FPW application is harnesses which interconnect gimballed devices such as stable platforms, gyroscopes, or tracking mechanisms. FPW, formed hairspringlike into a spiral coil whose center is coincident with the center of instrument rotation, eliminates slip rings and provides nearly torque-free, electrically transparent interconnection between frame and moving element.

One of the earliest examples of preformed FPW was a "yo-yo" coil (trademarked *Retrax* by Sanders Associates). Built up with thermoplastic insulation, the FPW was wound into a bifilar configuration (both ends of the circuit extending to the outside of the coil) which was heated to the dielectric softening temperature and chilled, thereby locked in the coiled shape. The fixed end of the circuit was attached to the frame of an instrument rack; the other, moving end terminated in a connector mounted in the back of a drawer. When the drawer was pulled out for service, the coil of FPW unwound to provide service loopage; it rewound—because of stored plastic memory in the dielectric—as the drawer was returned. Another example of retractable cable is shown in Fig. 5-1. This 50-ft-long, reel-mounted, multiconductor FPW powered a submarine periscope.

Environmental

FPW and wire harnessing have mixed strengths and weaknesses in environmental testing. Wire harnesses are vulnerable at terminations, because these are prone to stress concentration failure unless potted or strain-relieved, costly procedures which are not required in FPW. The weak link in FPW is the adhesive used to join dielectric and conductive layers. These adhesives are the poorest performers in thermal and chemical exposure and usually are flammable unless specially formulated.

Electrical performance of FPW is determined mostly by the adhesive layer because it is in direct contact with conductors; it is incorrect to assume, for example, that FPW insulated with adhesive-bonded Kapton or FEP-Teflon will survive thermal stresses as well as Kapton- or Teflon-insulated wire. Since FPW adhesives are modified for increased flexibility and reduced lamination flow, the chemical resistance and electrical performance of the base polymer are compromised, resulting in reduced insulation resistance and higher dielectric constant.

Assuming the same primary insulation and conductors, the generalized performance comparisons are

- Because of its lighter weight, FPW has better shock and vibration resistance. Specifics of the application (particularly mounting, strain relief, and support) strongly influence results, but in general FPW is superior. FPW is the interconnect technique used in fuses and other cannon-launched projectiles which must withstand extreme (above $1000g$) acceleration. See Fig. 5-3.

- Conventional adhesive-bonded FPW cannot survive sustained temperatures much above 125°C, or cycling humidity/temperature, because the conductor/adhesive interface will degrade. Wire constructed with the same primary dielectric but no adhesive can withstand higher temperatures. FPW manufactured with high-temperature adhesives or by fusion bonding utilizing fluorocarbons and stable conductor surface treatments provides significantly improved elevated temperature performance but at higher cost and with limited availability.

- Low temperature exposure is not detrimental to FPW, provided the exposure is not combined with flexure or vibration. The unitized structure of FPW is stiffened by exposure to temperatures below $-25°C$, but use in liquid nitrogen is common; wire harnesses are not as affected, assuming similar dielectrics, because the structure isn't as rigid—the dielectric can slide over the conductor bundles to distribute local stress.

- Combined elevated temperature and severe folding or creasing may cause creep or long-term flow of FPW adhesives, leading to permanent wrinkles or creases and potential conductor exposure. Relatively mild temperatures—80°C—are sufficient to provoke enough plasticity to cause concern. Wire cannot be formed as severely but won't fail from adhesive movement because there is no adhesive: the dielectric surrounds each run as a tube or sleeve. Good design allows FPW installation without creasing.

- Dielectric strength in FPW is very high; it can be expected to exceed 1000 V per mil of dielectric thickness since it is a sum of the dielectric strength of adhesive plus film. As most modern packages operate at semiconductor voltage levels, which seldom exceed 100 V, dielectric breakdown is very rare. Insulation resistance may be problematic with elevated temperature exposure, even under low-impedance semiconductor conditions, because flexibilized FPW adhesives lose their insulating properties as temperature rises. Since wire interconnects do not contain adhesives, inherently do include some (variable) degree of air separation, and have greater dielectric thicknesses, insulation resistance failure in wire harnesses is very uncommon.

- Flexural endurance is a particular strength of FPW, a natural result of form factor. For extremely long life requirements—hard drive flying head circuitry is a good example—relatively unplasticized adhesives and stiff dielectrics are used with thinner foils and balanced cross sections for maximum flex life.
- It is possible to manufacture wire with a tightly bonded dielectric—so-called watertight construction as found in cables for underground or underwater use—but at elevated cost. The manufacture of FPW tends to produce a tight seal between conductor and dielectric without added cost. Where FPW is exposed to pressure differentials, this characteristic is valuable but should not be assumed: special attention to conductor preparation and lamination is required for the best seal. In general, FPW is easier to gasket onto as an entity and far superior to conventional wire, conductor to conductor, for pressure differential use. See Fig. 4-20 for FPW with an integral gasket.

Troubleshooting, Repair

Correction of a nonfunctioning electronic system can be divided into two steps: repair of interconnect and repair of circuit.

A major advantage—and disadvantage—of wire harnessing is universal variability. Each harness can be made different from all others and, unfortunately, may be different even when sameness is the goal. If analysis shows that a wire harness has a missing or broken interconnection, it can be reworked and repaired by removing and replacing damaged conductors, thereby restoring original appearance and performance.

FPW is factory-produced as a system of unitized conductors, and hence is more difficult to repair. If an FPW conductor is broken or the dielectric is damaged by an overload, the failed section must be cut away, the dielectric must be scraped from the ends of the affected conductors, and a replacement or bypass conductor attached by solder or conductive adhesive to the cut-off conductor ends. The repair conductor with its attachments can be coated with insulating tape or potting compound to complete the fix. Alternatively, a jumper wire can be soldered to the end points of the desired circuit and routed over the outside of the FPW. Either is a comparatively cumbersome, if functional, process.

Rework for engineering change is similarly easier in wire harnessing and more difficult with FPW.

Because FPW is volume-produced in transparent dielectrics with conductor positions that do not change, identifying and examining the voltages and current flow on each conductor run for circuit analysis is much simpler than with wire harnessing. Wire harnesses can most easily be probed for trouble isolation at node points where conductor identification is possible.

Summary

FPW is the best interconnecting method for densely interconnected, volume-produced electronic assemblies. The start-up costs and delays are significant, but lowered assembly cost, stability and consistency of performance, error-free assembly, low weight, compactness, and esthetically pleasing appearance are the paybacks.

Wire is best for low-density or long-length interconnects, low production volumes, and high-rework applications.

Chapter 3

Manufacture of Flexible Printed Wiring and Printed Wiring Boards

Introduction

Recent trends toward products such as PCMCIA and "smart cards" have brought printed wiring board (PWB) technology closer to flexible printed wiring (FPW). The 0.002- and 0.003-in dielectrics used in these products are less stable and harder to handle than traditional PWB materials, inflicting FPW-like manufacturing losses. At the same time, the FPW industry is enjoying the emergence of a new class of materials called *adhesiveless* laminates—composites of dielectric and foil which are even thinner, yet more stable and predictable. If these trends continue, PWB and FPW manufacture will converge with rising demand for high-density multilayer products and acceptance of the new materials. For designers and purchasers of the '90s, it's correct to state that PWB and FPW manufacturing, superficially similar, are quite different.

Comparison of PWB and FPW Manufacturing

The printed wiring board is a common product found in all sorts of electronic equipment, from computer mainframes to TVs and electronic toys. FPW is less well-known; it is typically buried deep inside complex packages which are opened only by skilled technicians. Individual process steps used for FPW closely parallel those used to produce PWBs, and much of the equipment, chemistry, terminology, and engineering is common to both products, leading to a dangerously misleading impression of similarity. The danger is particularly acute for PWB-familiar designers and purchasers who assume that they're equipped to deal with FPW.

There are four major materials-related differences between FPW and PWB manufacturing. Two appear in the end item, two affect manufacturing process, and all reduce manufacturability and increase cost. It's important to review these differences and clear up misconceptions, because a little bit of knowledge is dangerous. The differences are:

Transparency

Dimensional stability

Handling

Coverlayer/covercoat

Transparency, Stability

Transparency

The function of PWB is to physically support and interconnect components. As a result, PWB laminates must have considerable mechanical strength in addition to good dielectric properties. They are generally composites of thermosetting resins reinforced by a fiber matrix. Examples are epoxy or polyimide resins with glass cloth reinforcement, polyester resin with polyester fiber mat support, and phenolic resin with cellulose paper. The combination of thermosetting resin with heat-resistant fiber produces a structure which withstands in-process thermal and mechanical stresses with relatively little distortion. An important unintended consequence is that PWB substrates are opaque.

FPW is a flexible product with quite different material requirements. Its function is to interconnect in three dimensions and between moving parts: in this use, structure and rigidity are undesirable. The best dielectric for FPW is an oriented polymer film, good examples of which are the currently popular polyimides and polyesters. These are transparent films; FPW has traditionally been built in transparent materials to the point where customers regard opaque dielectrics (fire-retardant systems are an exception) as suspicious. While polymer films have more than adequate mechanical properties to protect FPW conductors once they're built into electronic assemblies, these substrates are affected to a greater degree by in-process manufacturing stresses than are more robust PWB laminates.

Transparent dielectrics significantly influence inspection strategy. Both PWBs and FPW are graded by mechanical means (dimensions) and inspection of the surfaces. Because the transparent dielectrics allow it, the internal volume of FPW is also visually inspected—frequently under 10× magnification! Review of FPW inspection records will show that foreign material is a major reject cause. It is not their fault, and not a denigration of PWB manufacturers, but it is nevertheless true that much FPW is rejected for foreign material which wouldn't be visible in a PWB.

Visual inspection and rejection of FPW occurs throughout the manufacturing process and continues even at customer facilities: there's no statute of limitations. Nothing is hidden, because the internal volume of FPW is always visible. Once a hair or fiber is discovered embedded in an FPW dielectric, there's no rework and no recourse; standard practice is to reject because it's safest. Nor is it necessary to actually have something "foreign" there: change in appearance is sufficient cause for rejection. For example, copper is a reactive metal which easily discolors in contact with even low concentrations of contaminants. Unless the source and composition of off-color conductor surface areas can be proven, the product is rejected.

Selling price is affected by yield. If all FPW rejected because of foreign material found solely by visual inspection—unidentified, never touched, but not conductive—could be retrieved, the benefit in selling price and profit would be enormous. PWB has never been as inspectable and benefits thereby in more predictable production flow, higher yield, and better profit.

Dimensional stability

Dimensional stability—or, more accurately, material *in*stability—affects yield throughout the FPW manufacturing process. The Institute for Interconnecting and Packaging Electronic Circuits (IPC) dimensional stability test consists of measuring a length along both axes of a laminate, then etching all the foil cladding away and remeasuring; the change is a measure of stability.

FPW materials usually shrink after etch; typical values are on the order of 0.1%—about 1000 ppm. This seems like a small change, but at this stability an etched pattern from 10-in artwork is 9.990 in long; if you wanted an 18-in circuit, it's now 17.982 in, with more processing and more change to come.

Each thermal exposure in the production sequence may produce changed dimensions as the dielectric is softened and stressed. The reinforcing fiber matrix in PWB materials is a stable backbone which protects against this effect. Examples of FPW thermal processes which cause shrinkage or distortion (defined as uneven or unpredictable movement) are coverlay and multilayer lamination and plasma-process and oven-bake cycles used to cure nomenclature marking, drive out residual moisture, prepare for soldering, and so forth.

But loss of overall length isn't the most serious consequence of dimensional instability: FPW is always designed with excess length to compensate for tolerance buildup and to provide loopage for assembly and service. The major impact of instability is on tooling and panel size.

PWB is produced in immense panels—24 × 36 in is not uncommon—for maximum efficiency and least labor per circuit. FPW can't be run in large panels with multicavity tooling because material instability upsets tool alignment. A common FPW panel size is 18 × 24 in; in this panel, distance from tooling hole to circuit could be as great as 9 in. Assuming 0.1% shrinkage, that 9-in dimension becomes 8.991, a shift of .009 in. Since a common tolerance for alignment of hole to pad is 0.015 in, and pinning has an additional tolerance of 0.002 in, it's obvious that shrinkage in FPW materials impacts precision, thus yield.

Tactics to compensate for instability include:

1. Reducing panel and tool size
2. Factoring (expanding) artwork by expected shrinkage
3. Using postetch, autoalignment schemes (see Fig. 3-1)
4. Cluster registration (multiple datums)
5. Leaving as much copper as possible in the circuit

Methods 1, 3, and 4 improve tool alignment but add to cost without improving value.

Artwork factoring is a powerful but underutilized technique for nullifying material shrinkage. Here's how it works: Shrinkage is the result of stress in the laminate. For example, if dielectric film and foil were bonded together at 180°C and shrinkage measurement (and tool alignment) occur at 20°C, the temperature difference is 160°C; if the film has a coefficient of temperature expansion (CTE) of 30 ppm and copper has a CTE of 17 ppm, theoretical shrinkage is 2080 ppm, or about 0.2%.

Provided that, for the duration of the production run, the selected film continues to have the same CTE, is produced with consistent residual stress, and the lamination conditions aren't changed, the laminate should continue to show 0.2% shrinkage, which can be offset by enlarging the artwork by the same amount, an easy task for modern computer-aided design/manufacturing (CAD/CAM) systems.

Shrinkage is determined by the amount of remaining copper in the circuit design, and can be quite different along one axis compared to the other, factors which complicate artwork compensation. The best procedure is to apply a nominal 0.1% factor on both axes, etch a test

Manufacture of FPW and PWBs   23

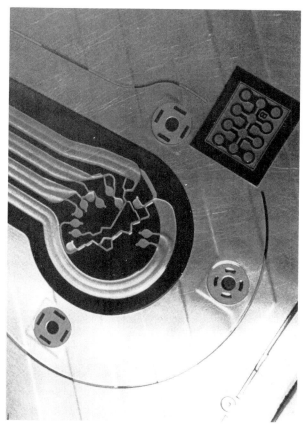

Figure 3-1 Close-up of cluster registration. Note three optical fiduciary patterns which have been automatically centered and punched; note also shielded circuitry and ventilation pattern visible in the embossed foil. Rectangular pattern with the nomenclature B is a test coupon. (*Courtesy Teledyne Electronic Technologies.*)

lot of panels, remeasure, and adjust if necessary. However, matching the artwork factor to shrinkage is a statistical problem which may require many measurements to determine the best factor for centering artwork size on average material movement.

Designers should appreciate the issues raised by conflict between artwork dimensions and drawing dimensions. An all-too-common FPW manufacturing headache is resolving disagreements which are embedded in a design data package: which governs—artwork, concentricity, or location?

Handling

PWB and FPW manufacturing technology must take into account the inherent nature of the materials. PWB laminates are stiff and strong, easily processed even with automated loaders and stackers, while lack of structure in FPW laminates makes them hard to move, even by hand, without damage.

It's easy to talk about 0.002-in films clad with 0.0005-in foils: conceptually you can visualize a proportionate cross section with etched runs, throughholes, and coverlays. But handling such materials—the simple process of picking up an 18- × 24-in sheet and placing it on a conveyor—is a much harder task. In the real world picking up an FPW laminate wrinkles and creases it because the foil cladding is easily stretched, leaving a distorted surface and dimensionally upset dielectric. What was a smooth, flat sheet in its receiving box is now something very different: photoresist won't bond to this irregular surface with an etchant-tight grip, nor will postetch pattern size be as expected. In a nutshell, this is a major FPW manufacturing drawback: FPW laminates require careful handling.

The effects of handling damage are subtle; they appear as elevated scrap losses categorized under a variety of other headings. Like dimensional stability, the effect of material handling is on factory cost, not product value. The FPW manufacturer deals with this problem by several techniques including:

1. Sliding, not lifting, wherever possible
2. Racking
3. Reducing panel size
4. Processing roll to roll wherever possible
5. Packaging and transporting in groups (see Fig. 3-2)
6. Using leaders and trailers
7. Processing back to back

Leaders are required to get FPW laminates through entry and exit seals in conveyerized production equipment. Such equipment will pass PWB laminates without special care because they are stiff enough to raise the rollers or push through the baffles. The FPW technique is to tape $1/16$-in PWB laminate to the end of the FPW panel, pass the combination through the equipment, then untape and reuse the leader (see Figs. 3-3 and 3-4). Added labor and material is expended, particularly compared with a PWB process using automated load/unload equipment, because an operator is required at both ends of the equipment.

Manufacture of FPW and PWBs 25

Figure 3-2 View of carrier trays and rack. These shallow handling aids protect and organize multiple sheets of flexible circuitry as they move from operation to operation. (*Courtesy Teledyne Electronic Technologies.*)

Figure 3-3 Detail of leader taped to flexible sheet. (*Courtesy Teledyne Electronic Technologies.*)

Figure 3-4 Detail of conveyor showing top and bottom rollers and spray bars. (*Courtesy Teledyne Electronic Technologies.*)

Back-to-back processing is a method for reducing handling damage which also lowers process labor in single-sided FPW manufacture. In this technique, two sheets of FPW laminate are temporarily assembled to a central carrier sheet with only enough adhesion to allow processing, then separated at the end of the sequence. Because PWB and FPW equipment is designed for double-sided processing, the back-to-back technique allows simultaneous treatment of two single-sided circuits, for reduced labor cost. This method can't be easily used with predrilled laminates, which may leak, trapping process chemicals and contaminating other baths and the product. However, peripheral tooling holes can be used provided the two sheets are offset from the carrier. See Fig. 3-5.

Examples of successful self-adhering carrier materials are vinyl films and epoxy-glass panels. Hard tempered aluminum foils (drill entry sheets) make excellent carriers; they require narrow strips of adhesive around their perimeter which, after the process, are cut off to separate the panels. An adhesive/foil system is used where greatest security is needed.

Influence of foil choice on handling

FPW laminates are traditionally based on rolled-annealed (RA) foils (see Chap. 7, "Conductive Materials"). This choice is traceable to early unhappy experience with FPW built with ED foils which lacked good

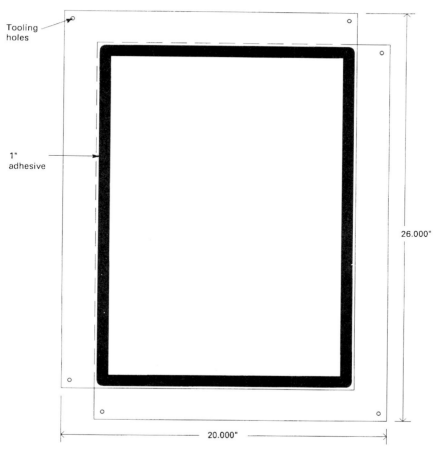

Figure 3-5 Offset carrier. Illustrates method for temporarily securing two sheets of FPW laminate to a stiff carrier in an arrangement which provides access to tooling holes in the perimeter.

flexural properties. ED (electrodeposited) foil is less expensive in FPW thicknesses and was readily available, compared with RA, at that time. However, a rash of fractured conductors convinced FPW engineers to pay the extra cost for RA foils, which remain today the preferred choice. But RA foil is much more susceptible to handling damage. RA can be permanently creased and dented with incredible ease, and each fold or buckle, even from light mechanical cleaning, constitutes a stress concentration which may lead to unexpected material movement after etch. Susceptibility to handling damage in FPW laminates is thus exacerbated by foil choice.

Use of RA foil for flexible materials is no longer logical, since the modern ED process as practiced in so-called adhesiveless materials (see Chap. 8, "Adhesiveless Materials") produces thin copper foils with superior flexural properties and greater resistance to handling damage, at lower cost.

Coverlayers

PWB is a planar product. It can't short-circuit to itself. In many applications PWB conductors are left uncovered for low cost and easy assembly while in other instances a simple coating of liquid dielectric is applied to protect the circuitry from dirt and solder. In contrast, because it is folded, bent, and creased in installation and service, FPW almost always requires fully insulated or coverlayed conductors.

The coverlayer process consists of bonding a second composite of adhesive and polymer film, usually identical to the base system, onto the etched conductor runs. This process essentially doubles FPW material content, increases process labor, and complicates the production of termination areas which must be free of coverlay. For detailed treatment of coverlayer/covercoat processes see Chap. 8, "Manufacturing Processes."

Summary

Significant differences between FPW and PWB manufacturing procedures impact design and cost. Compared with more familiar PWB processes, FPW production methods require more labor and materials yet suffer reduced yield because of material instability, susceptibility to handling damage, increased inspectability, and use of coverlayers. Effective FPW designs must take these factors into account, and evaluation of FPW quotations without consideration of the added complexity and cost which they inflict will result in erroneous conclusions.

Chapter 4

Design

Introduction

Because flexible printed wiring (FPW) is a custom product, manufacturing can't begin before an enormous amount of up-front design and custom tooling is completed. It's a tough job which involves making tradeoffs between conflicting mechanical and electrical issues, buyer and seller, and cost and delivery requirements. It's made even harder by the fact that FPW design is like a photographic negative: it comes first, and once fixed, determines everything that follows.

The FPW designer must function for all interests—product, assembly, and quality—while developing a comprehensive set of cost-effective tools, manufacturing instructions, and quality requirements. His task includes determining the right materials, inspection methods, quality standards, and artwork to produce the desired interconnections, in a package put together with full appreciation for design for manufacturing, design for assembly, and design for test. Design is not an easy task: it requires considerable experience and judgment and recognition by management that design is much more than drawing a few lines on a CAD station.

There is a sizable library of industry-accepted specifications governing FPW and its assembly, more than enough for thorough control of materials and process. Designers should make maximum use of them, but select the most open tolerances and standards possible without compromising performance and quality.

Assembly technique—soldering, pressure connection, crimp—has profound impact on FPW design. Conservative, military-oriented design requirements and assembly methods drive cost upward while commercial, results-oriented techniques lower cost. The termination technique determines material choice and terminal pad design; pad design drives product design because terminal areas are the most dense and therefore define the number of conductors in a layer. Choice of dielectric and number of layers are fundamental elements of cost. For these reasons, the designer is urged to review the impact of the assembly method on FPW design with engineering and quality parties before starting the design process.

Data Requirements

Information that's required:

1. Complete mechanical descriptions and dimensions of all termination patterns and intervening structures
2. Electrical interconnections including current loads, resistivity, and shielding requirements
3. Assembly and servicing sequences; access requirements
4. Environmental/test requirements; test conditions
5. Documentation/certification
6. Cost target

The magnitude of the design task should not be underestimated from this short list. FPW design is similar to, but orders of magnitude more difficult than, PWB design: to begin with, FPW involves three dimensions as well as dynamic flexing, neither of which figure in PWB design. Cost efficiency requires packing the maximum number of conductors into the least area, a tough task given the typically complicated outlines found in FPW. In addition, as detailed in Chap. 3, there are major manufacturing differences between FPW and PWB which further complicate the picture, i.e., coverlayers, transparency, and dimensional instability. An excellent, detailed guide to FPW design is IPC document D-249, "Design Standard for Flexible Single- and Double-sided Printed Boards," published by the Institute for Interconnecting and Packaging Electronic Circuits (IPC).

Intellectual Approach

The first step is to get oriented to the overall picture, then to establish a layout which accommodates the conductors and assembly procedures and a design/quality approach appropriate to the project, be it a high-volume commercial application never to be serviced; a military/aerospace item, man-rated and documented to the maximum; a totally new, developmental product intended for automated assembly; or a cost-reduction retrofit.

Expected environmental conditions should be carefully considered.

It's unlikely that conditions inside electronic equipment will be as severe (in elevated temperature and pressures) as those FPW experiences in manufacturing and assembly processes, but flexibility at low temperatures and shock resistance may influence material choice or dictate special mounting provisions.

Keep in mind that two primary reasons for using FPW are to lower assembled cost and improve performance. The ultimate FPW design is a series of parallel conductor runs filling a rectangular envelope; this provides highest conductor density for lowest cost-per-run and best panel utilization. Real applications have curves and corners which reduce packing efficiency of conductors into the FPW outline and of multiple copies of FPW into the panel composite. Between termination areas FPW design can approach theoretical conductor density and design efficiency; termination areas always require increased conductor center-to-center spacing to accommodate pads, which reduces efficiency. Conceptually, strive for straight layouts with every conductor routed directly to the correct terminal pad, but expect routing problems.

Procedure

Mock-ups

Make a dimensionally accurate mechanical mock-up. Identify all termination points, using the same connector designations and pin identifications as on the run list. Be sure to include everything which must be avoided or accessible—fan intakes, adjustment screws, and so forth. Get clearly in mind the intended sequence of assembling components to FPW to package because it affects configuration; a circuit designed to be subassembled to connectors then installed in a box isn't the best layout if the FPW has to be assembled to connectors inside the package. Also take into account field repair or servicing needs—if hardware must be field-replaceable, it shouldn't be buried under layers of FPW.

Circuit analysis

Analyze the wiring list and reassemble it by circuit type and geography. Example circuit types are high-current, sensitive, shielded, and controlled-impedance. The geographical arrangement will identify interconnection patterns: where circuits start and where they go. These patterns set the overall FPW design. Autorouting programs in CAD software can create artwork designs directly from a wiring list, but are blind to the benefits of organizing conductors by circuit type, i.e., more efficient use of shielding, better isolation of sensitive runs, selective use of heavier foils for power runs, and so forth. Use the computer where it's good, use common sense as well; the combination will result in a more efficient design.

Design rules

Establish design rules for:

1. Conductor cross sections based on current loading or resistivity requirements
2. Conductor-to-conductor spacing
3. Terminations: minimum annular soldering land, connector contacts and surface finish, plated throughhole (PTH) if used
4. Edge distance
5. Test points, nomenclature, other nonfunctional items

It's very important to demand realistic electrical specifications: be certain that requirements for each conductor run are based on engineering analysis rather than historic practice or guesswork. Conductor size is a basic element in FPW design and strongly affects circuit cost, thus should be carefully considered at the outset (Fig. 4-1).

Temperature rise is strongly influenced by insulation thickness, number of energized conductors, and specifics of package design, including air flow. Rectangular FPW conductors can carry greater current at the same cross sectional area as round wire because they have greater surface area and therefore dissipate heat more efficiently.

There is very little current capacity information available for copper less than 0.0014 in (traditional "1 oz") thick, but thinner foils have

Width, in	0.0014 in thick		0.0028 in thick	
	Capacity,* A	Resistance, mΩ/ft	Capacity,* A	Resistance, mΩ/ft
0.005	0.25	1200	Impractical	600
0.01	0.37	600	0.65	300
0.015	0.5	400	0.8	200
0.02	0.7	300	1	150
0.025	0.76	240	1.2	120
0.05	1.4	120	2.1	60
0.1	2	60	3.45	30

*Current capacity is from MIL-STD 2118 and is conservative for a single energized run.

Figure 4-1 Current-carrying capacity and resistivity for FPW conductors in copper. Assumes 10°C rise and full insulation.

proportionately more surface area therefore higher relative current capacity. Thinner foils are preferred in FPW for additional reasons:

1. Etch precision goes up with thinner foils. Finer lines are possible, with less variation, at lower cost.
2. Coverlayering is more precise because less adhesive is needed to fill thinner etched patterns, therefore unwanted adhesive flow into apertures is reduced.
3. Flexural endurance improves with reduced foil thickness; increased life is inversely proportional to the square of the thickness; e.g., 0.0007-in foils have (at least) 4 times the flex life of 0.0014-in foils.
4. Adhesiveless materials (see Chap. 7) are available with foil claddings down to 0.0001 in, allowing production of 0.0003-in lines and spaces with liquid photoresist.

Resistivity for other thicknesses can be calculated from cross sectional area from the formula

$$WTR = 6000$$

where W = width, mils
T = thickness, oz
R = resistance, mΩ/ft

For example, a run 0.01 in wide (10 mils) and 0.0007 in thick ($\frac{1}{2}$ oz) will show resistivity of 1200 mΩ/ft. Resistivity in other alloys requires further adjustment based on specific resistivity:

$$R_{alloy} = R_{copper} \times \frac{\text{Resistivity}_{alloy}}{\text{Resistivity}_{copper}}$$

Design rules for conductor width and spacing should take into account:

- Minimum conductor width required for current-carrying capacity or conductivity
- Spacing (and edge distances) adequate for circuit voltage
- Manufacturability/cost issues (wider is better)
- Etch factor
- Safety factors

Etch factor—the added width built into the artwork to compensate for etch losses—is related to etch chemistry and etch process control. A conservative allowance is 0.002 in per 0.0014 in of foil thickness.

Safety factors should be as large as possible, consistent with cost. It can be economically reasonable to eliminate layers by lowering the safety factor—setting narrower width and spacing rules, thus increasing conductor density. See the discussion of conductor resistivity in the next section, "Preliminary Layout." Short necked-down sections of conductor have virtually no effect on overall resistivity or current-carrying capacity. Good designs show consideration for the tradeoff of producibility with feature size and density.

Dielectric breakdown and short circuits to ground should be considered, but FPW dielectrics, which have dielectric strengths measured in hundreds of volts per mil, are seldom challenged by modern circuit voltages. The least manufacturable spacing and film thickness will be more than enough for ordinary circuit use.

Always leave as much copper as possible in FPW designs: it costs money to remove it, and dimensional stability improves with increasing copper content. Copper can be left inside the FPW and between pattern repeats in a composite. In circuits, the benefits are

Extra copper helps maintain preformed shape

Copper reinforcement around inside corners and mounting holes adds greatly to tear resistance (at no added cost)

Grounded copper planes improve electrical isolation

In panels, these are the benefits:

Copper borders improve dimensional stability

Added copper stiffens the panel, reducing handling damage

Figure 4-2 lists standards and reduced-production (extra high quality) dimensions and their tolerances, as well as tolerance levels. A design which can be built to standard tolerances and level A precision will be lowest in cost.

Preliminary layout

Put it all together—the mechanical mock-up, design rules, circuit type, and geographic interconnect patterns—and make "paper doll" mock-ups which:

- Are wide enough between terminal areas to accommodate the number and size of conductors required.
- Fit neatly on the mechanical mock-up with reasonable allowance for service and installation. Factoring the number of circuits routed from area A to area B times conductor width and spacing as determined by electrical requirements calculates the minimum width of

Feature	Standard	Tolerance	Reduced Production	Tolerance
Hole size	>0.015	±0.002	>0.010	0.001
Conductor width	0.008	0.002	0.005	0.001
Spaces	0.005	0.002	0.004	0.001
Hole-to-pin clearance, diametral	0.01			
Minimum annular land	0.015		0.008	
Edge distance:				
Steel rule	>0.02	0.01	0.01	0.007
Matched die	>0.010	0.005	0.008	0.003

Dimensional tolerances	Length	
Level	<12	<18
A	0.034	0.04
B	0.022	0.024
C	0.012	0.018

Figure 4-2 Manufacturing tolerance chart. All dimensions in inches.

FPW. Width of choke points or bottlenecks in the mechanical mock-up determines the maximum number of conductors per layer, thus the number of layers required. A little trick here: resistivity is a function of average width. If the bottleneck is short, conductor size can be reduced to allow more runs per layer (and fewer layers) with added width in other areas to restore conductivity. Be aware that when multiple layers of FPW are bent around a radius, outer layers must be longer to compensate for greater path length. This is called *progression* and is the basis for "bookbinder design." See Fig. 4-3 and Figs. 9-11 and 9-12.

- Have all contact points in a circuit in one piece of FPW.
- Cluster like circuits with like circuits—high current runs with high current runs, sensitive runs with sensitive runs, etc.

Explore and optimize overall layout, fold-overs, progression design, hinging, assembly, and service access until a good shape is established. Make copies and nest them together to determine how many circuits will fit into a production panel as a cost estimator. Modify and optimize.

Effective designs have high conductor density; expect to find that connector areas are the bottlenecks. It's a good idea to start the development by estimating the number of runs which can terminate at a

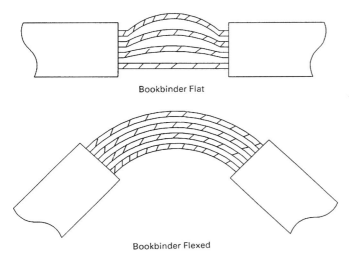

Figure 4-3 Progression, or bookbinder, FPW design to provide unbuckled FPW layers in a tightly flexed RF circuit.

connector in each layer, then size the paper dolls based on that number.

Getting conductors in the right sequence to gracefully flow into dense areas is a major design headache. In rare instances, the designer will have *freedom of pin address,* i.e., will be allowed to determine the pin-to-pin interconnection pattern based on natural sequences that the layout dictates. In the real world, FPW design comes late in the program, after the wiring plan is fixed, with little time for rearrangement. Consequently, there will be many instances where the left-hand conductor goes to the right-hand pin and so forth. PTH is a universal fix for this problem; it allows rearrangement by interlayer jumpering. See Fig. 4-4. Methods available to the non-PTH FPW designer include foldouts, reversed address, double-joint jumpers, and feedthroughs. See Fig. 4-5.

One-piece FPW designs are elegant from a design point of view but proportionately more costly, per conductor, to build; multiple smaller pieces nest more efficiently onto FPW panels and have higher production yield; they are therefore cheaper to procure. It's important to note that if assembly technique and quality standards allow multiple solder joints at a terminal, designs may be simplified by breaking up complex circuits into separate FPWs which are rejoined by assembly to common terminals. Smaller FPW pieces aid in grouping by circuit type and make it possible to assemble and test in smaller units, further reducing cost.

Design   37

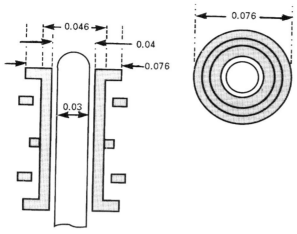

Figure 4-4 The PTH technique is a powerful method for dense connector termination; it easily brings multiple layers of FPW to the pattern while providing a single soldering surface for quick, reliable joining. But the PTH technique demands careful attention to tolerances and annular allowances which quickly consume surprising amounts of real estate.

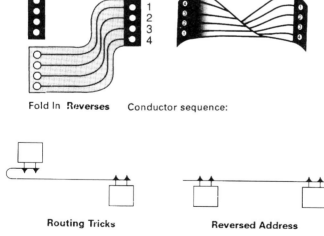

Figure 4-5 Reversal address. The sequence of conductors at a connector terminal pattern may be reversed by using a 180° bend, or fold-over, to reverse the direction of approach.

Refinements

Terminations

Terminals are the weak link in FPW and deserve a lot of design attention. They're exposed to extreme heat and/or pressure during assembly, if soldering is the technique, and high stress concentration throughout the life of the package regardless of assembly method because this is the point of transition from flexible FPW to rigid hardware.

The simplest and lowest-cost terminal is an enlarged area or pad on a conductor run which aligns with a connector pin or component terminal, engages it by means of a throughhole, and is permanently soldered in place through an aperture in the coverlayer or covercoat. This is the most common termination design for low-cost, high-volume FPW.

Data requirements. Design definition and documentation of terminals includes coordinated artwork, drill programs, and/or die design for:

- Terminals (artwork)
- Throughholes (drill programs or die design)
- Apertures (drill or route programs or die design)

Fold-over design or other techniques which increase routing area or reduce circuit layout area may force termination to pads from the back or reverse side. Reverse baring (see Chap. 8) increases circuit cost; designers must decide the net benefit of reduced area against increased manufacturing cost.

Terminals should be grouped together and aligned for strength in unity; each reinforces the other, protecting against stress concentrations. When only a few pads are present in a terminal area, particularly when the FPW approaches at an angle, strain relief or stiffening to distribute the stress and protect the pad-run interface is needed. See Fig. 4-6.

Tin-lead soldering is the dominant assembly technique, but pressure connections, conductive adhesive, and wire bonding are also used and gaining in popularity because they allow lower-cost dielectrics. When direct solder attachment is used, the bond between the pad and support dielectric is weakened by exposure to extreme temperatures and forces; larger pads with added ears or hold-down tabs help keep pads in place while they're soldered. Coverlayers which overlap onto terminations strengthen them considerably; it's good practice for coverlayers to overlap by .01 in, more if tabs aren't used. The junction of pad to run should always be filleted in the artwork design (see Fig. 4-7) to reduce stress concentrations at this vulnerable, highly stressed interface.

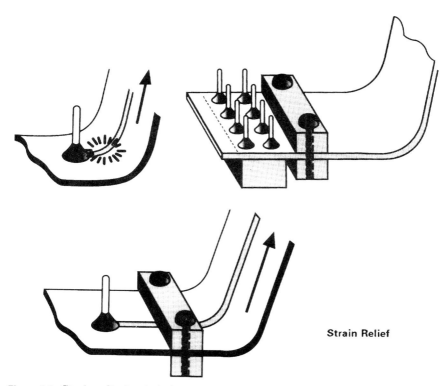

Figure 4-6 Strain-relieving technique.

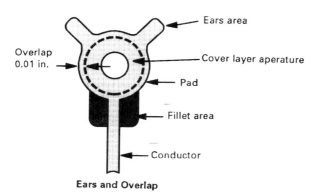

Figure 4-7 Ears and overlap. The junction of a terminal pad with its conductor should be smoothly blended or filleted in width on the artwork to avoid an abrupt change in cross-sectional area (and resulting stress concentration). Hold-down ears aid in securing the pad under the coverlayer. Coverlayer apertures should be sized to overlap the pads by 0.01 in or more.

Surface mount assembly. Direct attachment to FPW by surface mounting is an increasingly popular technique which leads to considerable package simplification. A special termination design problem, pads for surface mount assembly (SMA) must be configured for reflow solder attachment. SMA pads have no throughholes and are on close centers. Because the available volume of solder is limited, reliable reflow attachment requires careful consideration of pad size and shape and the thickness and aperture shape of coverlayers or covercoats. Further considerations are the wettability of the FPW dielectric and its effect on solder balls and postsolder cleaning. Much of the technology developed for SMA on printed wiring boards (PWBs) can be used with FPW, and the designer is urged to review this data carefully.

SMA guidelines:

1. Pads should be coverlayered or covercoated to prevent solder escape—apertures should define the best area for device attachment.
2. Pads should extend well underneath the coverlayer or covercoat to increase resistance to pad lifting from attach and rework cycles.
3. Coverlayer or covercoat thickness should be minimal to minimize impact on solder-cream deposition and device attachment. Because the coverlayer could prevent the device from settling into contact during reflow, it's possible that apertures should be larger than the component.
4. PWB device footprints should be used if appropriate. IPC D-249 contains a wealth of specific SMA information.

Connector approaches

Connector choice has important quality and systems implications and should be reviewed and decided on early in the design process.

More FPW conductors terminate at a connector than at all other types of hardware combined. The range of connectors which is available for solder (or pressure) attachment to FPW is immense and constantly increasing; it's fair to say that every style of connector is available with FPW-compatible backside contacts. Because terminal-pad density is at its maximum in connector patterns, this area deserves a great deal of attention: this is where the maximum number of conductors will be determined; this is where problems of pin address and conductor sequence will arise.

In cases where an FPW assembly mates with other systems, connector choice is dictated by the existing interface. In other situations the designer should choose a connector which has enough pins for all the interfacing circuits (plus spares), with enough spacing and routing room between pins to make the FPW design both practical and cost-effective. Excessive density is elegant but costly.

Most connectors are directly soldered to terminal pads, either by wave, reflow, or hand technique. Other joining techniques are pressure or crimp attachment and, simplest of all, direct interface in which the FPW terminal is gripped by a zero insertion force (ZIF) or low insertion force (LIF) mechanism or is pressed directly against a mating contact. Compatibility, reliability, cost, field service, density, and performance are all important connector topics; connectors are a major issue in FPW design.

Routing techniques. At connector patterns, each run flares out to a greater width to allow for throughholes with annular lands—0.015 in wide for simple FPW—thus consuming greater routing area. Multiple layers with a staggered approach (see Fig. 4-8), with or without clipped pins, are a straightforward method for addressing termination areas but require inspection after each layer is attached and may trap foreign material, forcing expensive rework. The *fold-in and fold-in reverse technique* (see Fig. 4-9) aids connector address but raises cost because it requires reverse baring and may reduce the number of circuits which can be packed onto a panel. Feedthrough (Fig. 4-10) and double-joint jumper (Fig. 4-11) methods are additional design techniques for reducing the number of layers. See Figs. 4-12, 4-13, and 4-14 for examples of FPW/connector interfaces.

Plated throughholes. There are many ways to approach a connector; the best choice depends on assembly technique, FPW area/nesting considerations, and quality requirements. The PTH technique is a specific fix for dense termination areas. It increases effective routing

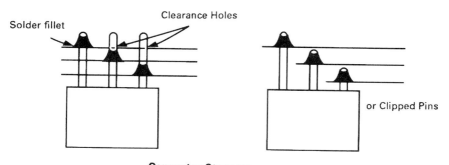

Figure 4-8 Connector staggering. Addressing dense connector patterns frequently requires multiple layers of FPW. Access to interior rows may be obtained either by providing clearance holes or by clipping previously terminated pins.

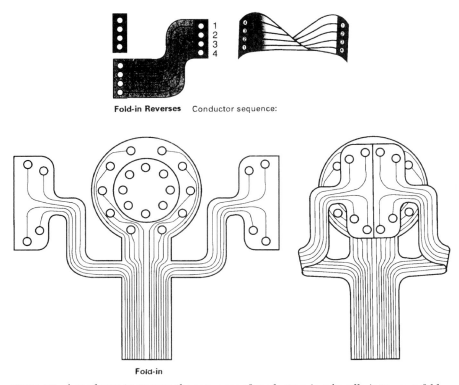

Figure 4-9 A good way to reverse the sequence of conductors in a bundle is to use a fold-in flap. This method also avoids space-consuming clearance holes, since the fold-in flap is in a different plane.

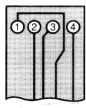

Figure 4-10 When the sequence of conductors in a group isn't correct for termination at a connector, a changed order can be created by feeding through the out-of-sequence runs and approaching the connector from the opposite site.

Feed Through

area because runs in other layers can be routed above or below the terminal lands. It aids assembly because all solder joints or connector contacts are in one plane, simplifying inspection and eliminating foreign material and contamination concerns found in multiple-layer methods.

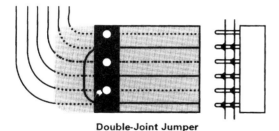

Double-Joint Jumper

Figure 4-11 Double-joint jumper. Double solder joints on unused connector pins can be employed to jumper out-of-sequence runs into correct positions in the pattern.

Figure 4-12 Fine-line FPW with two different right-angle connectors. Note neat, clean connector attachment. Bright circular areas are ground terminal pads for conductive silver shield coating (to be applied). (*Courtesy Parlex Corp.*)

The PTH process is expensive; estimated high-volume added cost per 18-in × 24-in panel for multilayer lamination and PTH process is $160. But remember that once PTH is used for connector patterns, it can be used anywhere else in the circuit design without added cost, since PTH is a per-panel process. (See "Dynamics" below.)

Tolerance stack-up for PTH shows the incredible rate at which real estate is consumed (see Fig. 4-4) by throughholes and their tolerances. Let's say the connector pin is 0.03 in in diameter. Assembly fit

Chapter Four

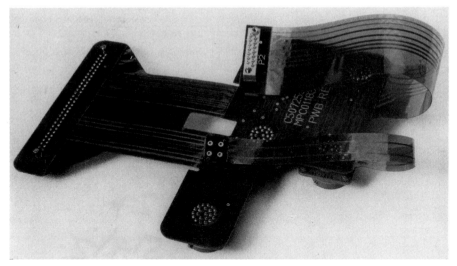

Figure 4-13 Rigid-flex assembly showing round MIL-style connectors, rectangular connectors, and PTH patterns used as connectors. Typical of military designs, this example has polyimide-glass "cap boards" supporting PTH terminations; this high-temperature dielectric system easily resists damage at solder attachment and coating to insulate the solder joints. (*Courtesy Parlex Corp.*)

and clearance to allow solder wetting forces the PTH to 0.04 in in diameter. Allowance for copper in the plated barrel adds 0.006 in to the diameter, thus minimum drilled throughhole size is 0.046 in.

According to MIL-STD 2118, the minimum annular land remaining after the hole is drilled must be 0.005 in on external lands and 0.002 in on internal lands. Thus minimum external pad size is 0.046 plus 0.01 in or 0.056 in, assuming a perfectly centered hole. (Internal lands may be reduced to 0.002 in but require 0.003 in allowance for etchback thus must be 0.056 in as well.)

Hole location is degraded by material shrinkage, registration, and drill machine error so a considerable fabrication tolerance—0.02 in minimum—is required. This adds directly, so our pad becomes 0.076 in in the etched form.

A reasonable etch factor of 0.002 in with a 0.004-in safety factor brings the artwork minimum pad to 0.082 in in diameter.

Tooling

Artwork is the principal output of the design process, but design also includes defining and documenting all production tooling which may

Design   45

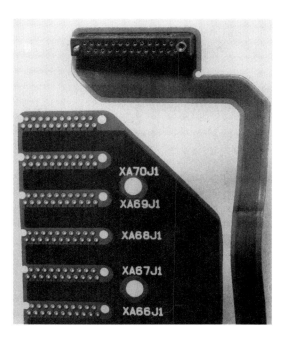

Figure 4-14 Complex, double-shielded FPW assembly (partially complete). Printed nomenclature clearly identifies each connector pattern. The assembly has drilled-hole tear stops and screw-attachment strain relief. The edge-on FPW approach with a 90° bend provides maximum adjustability in connector location for easy mating; a second 90° bend provides a flare-out area to allow wider conductor separation at terminals. (*Courtesy Parlex Corp.*)

be required. There can be quite a few tools involved, thus, compared with wire harnesses, FPW nonrecurring costs—most of which are for tooling—is sometimes very large indeed. To increase the importance of tooling, circuit production and test cannot begin without custom tools and the invested engineering effort to design them.

The number of tools, thus total cost, that is needed for efficient production depends on anticipated production quantity and required precision. Tooling must also be designed with an eye to available production equipment and, to some degree, the habits and normal procedures of the manufacturing organization.

Small or prototype runs can be produced with little more for tooling than artwork, which for this use can include center marks and cutting lines to define where hand-punched or drilled holes are needed and where to cut the outline with scissors or a hand knife. If the design proves to be adequate, and production quantity rises, labor required per circuit can be reduced and controlled as precision is increased by adding well-designed and -coordinated tooling to the basic artwork.

A general list of FPW tooling includes:

1. Artwork

2. Drill and routing programs
3. Outline and piercing dies
4. Lamination fixtures
5. Alignment fixtures (covercoat, stiffener, etc.)
6. Special handling or protective shipping devices
7. Electrical test fixtures
8. Forming and potting

All tooling which aligns to the circuit patterns, either as single circuits or in composite form, is conveniently generated from the CAD database. Because such databases include tooling hole locations, this practice assures maximum accuracy and agreement between tool and circuit.

Artwork

Nothing can proceed without artwork, which is the most critical tool and usually consumes the bulk of the design effort. This simple sheet of photographic film (or magnetic data for direct plotters) incorporates everything about the conductor pattern. It must be correct in every detail; it must include necessary compensations for etch losses and material shrinkage and, in composite form, also must make the best use of material and process capacity.

Drill and rout programs and dies

Coverlayer apertures are commonly generated by numerical control (NC) drilling, using a version of the terminal (or PTH) hole pattern which has the same locations but larger drill sizes as required.

Very high volume production, or unusual aperture shapes, may require piercing dies. Piercing is cleaner than drilling and doesn't require entry and exit materials. It also avoids the risk of heat-induced damage in adhesive-coated materials.

Routing is not easy to perform on FPW because the material isn't stiff enough to cut cleanly. Consequently, most FPW is outlined by use of a steel-rule die (SRD), a low-cost, NC-produced cutting tool which resembles a shaped knife edge embedded in a wooden block. Precision of SRD cutting is limited, however, and where close tolerances are required, machine SRDs or male-female or matched dies are used. See Fig. 8-22 in Chap. 8, "Manufacturing Processes," for typical die-cut tolerances.

Design 47

Lamination and alignment

Lamination and alignment fixtures are employed to assure alignment and to accommodate unusual thicknesses or to protect sensitive areas in lamination. These fixtures are ordinarily fashioned from aluminum jig plate and utilize tooling pins to achieve good precision without requiring high levels of operator skill.

Handling fixtures

FPW can be damaged during manufacture, assembly, or shipping. Support trays or mechanical mock-ups which secure large connectors and other components to a frame, relieving the FPW of mechanical stress, are often used as insurance against hidden damage. The range of shapes and materials which are used is as wide as the designer's imagination or experience.

Test fixtures

Test fixtures are readily constructed directly from the termination hole drill program. The program is used to create a matching set of holes in a dielectric sheet; spring-loaded contacts are pressed into the holes and align with the FPW terminations via tooling pins. See Figs. 8-23, 8-24, and 8-25.

Mechanical test fixtures for verifying specific shapes or locations are also sometimes provided with the production tooling package.

Forming and potting

Forming and potting fixtures are assembly aids. Forming, or creasing, usually references connector locations or locations on the outline; fixtures for forming FPW are pinned to tooling holes and present smooth, rounded edges at these locations to control the operation.

Potting fixtures are used to seal connectors or other clusters of terminations into a block of insulation to protect them from mechanical stress or accidental contact with other electrified circuits. See Figs. 11-3 and 11-8 for views of forming fixtures and refer to Chap. 11, "Assembly," for further details of potting.

Rigid Flex

The subjects of materials and manufacture are treated in Chaps. 7 and 9, respectively. Terminology is explained more fully in Chap. 9; read it and this section together.

Rigid-flex (RF) circuitry is the most complicated form of printed wiring. A designer looking at a typical (RF) circuit—thicker, rigid areas interconnected by thinner FPW areas in an uninhibited range of shapes and sizes—would be intimidated by the complexity (see Figs. 4-13, 4-22, and 4-23). Manufacture of the multilayered, mixed-dielectric products is a sort of postgraduate test of technology and process control, but their design isn't much more complicated than design of multilayered FPW—harder, but only in the volume of detail to be coordinated.

Conventional RF circuitry consists of outer rigid (PWB material) layers sandwiching a central stack-up of FPW layers. PTH termination is used to interconnect layers and bring all terminations to the outer surfaces of the rigid areas. These surfaces carry pads only, no conductor runs or other electrified areas because the purpose of the rigid layers is to protect all the circuitry from assembly abuse, short-circuiting, and damage.

All connector or hardware attachment occurs in rigid areas which, in addition to armoring the circuitry, provide strain relief and component support. Rigid areas are interconnected with each other by FPW layers which are individually coverlayered but normally unbonded to each other for better flexibility and reduced bending stress.

Design approach

RF is processed in panel form and only emerges as rigid and flexible areas after final routing. Until this stage, layer dielectric is incidental to the manufacturing process and the designer can consider all layers identically supported. Design proceeds as for any FPW—establish mechanical layout with accurate dimensions for all termination points, reassemble wiring list by similarity of circuit type and geography, determine overall FPW outline, set design rules, route the conductors—with these additions

- Outlines for both rigid and flexible areas
- Extra tooling for slots, windows, and fillers
- FPW area decisions: single or double sided; bonded or unbonded
- Materials choice (see Chap. 7)
- Specification

Mechanical layer design follows cluster tolerance procedures in that dimensions throughout a rigid area can be toleranced to NC machine levels (0.002 in), while lengths of FPW layers between rigid areas are fractionally toleranced. The layout should provide slightly eased (longer) FPW between rigid areas, but be on the lookout for buckling between multiple FPW layers in the flex areas at installation.

Avoid progression or bookbinder designs. Use adequate FPW length between rigid areas to accomplish the same stress reduction without added manufacturing complexity. Unbonded single-sided FPW layers are more flexible than double-sided FPW, at added manufacturing cost. They're usually preferred to bookbinder design, but query proposed vendors to determine their preference. Vendors specializing in the bookbinder technique may quote prototypes competitively, but be aware that bookbinder design limits the vendor base for future procurement.

Termination to bare FPW pads is discouraged. RF requires the PTH process, which can therefore be used to maximum advantage anywhere in the design without added cost. Bare pads complicate the RF manufacturing process, particularly if solder coating is used. This is because, in RF, the last manufacturing process should be final routing—past this point the circuit is too complicated and vulnerable and has too many surfaces and crevices, to be passed through wet processes other than cleaning. Solder-coated surfaces will degrade at lamination temperature, possibly becoming unsolderable; therefore solder should be applied after lamination. Since the bare pads will be sealed inside the panel for protection from plasma, electroless deposition, galvanic plating, and tin-lead reflow, they're accessible only after final routing. Tinning and cleaning, at this point, are hand operations to be avoided.

Alignment, composites

As with any multilayered product, the first priority is to provide for accurate registration and alignment between the layers. RF circuitry is a high-value-added product in which material cost is a minor element; packing as many RF circuits as possible into a composite is poor economics if it results in excessive distance between circuit and tooling pins. The best approach is to use the postetch-punch (PEP) technique—each layer has multiple optical targets used to create tooling holes in the etched, coverlayered part. Layers may be processed in panel form but should be cut apart, postetch-punched and laid up for multilayer lamination as individual layers to neutralize panel shrinkage. Layers are run, for example, four-up on an 18-$\times$24-in panel, then cut apart for layup in 9-$\times$12-in RF panels.

Rigid areas should be surrounded by tooling pins which engage the stabilizing copper border every 6 to 8 in. Cluster tolerancing—each rigid area a separate zone—should be used. Cost of additional tooling pins or custom lamination fixtures is justified.

Uniform rigid-area thickness is highly preferred. Rigid areas of different thicknesses can't be simultaneously PTH-processed and will force sequential lamination at considerably increased cost. If FPW interconnection to a rigid area isn't needed, the layer can be deleted from that area and replaced with prepreg to maintain uniform thickness.

PTH termination must be at least 0.125 in away from any outline; further if possible.

PTH process will be complicated by mixed rigid and flexible dielectrics; plasma treatment is mandatory. Expect difficulty achieving *three-sided lock* etchback (removal of dielectric from top, face, and bottom of pads at hole interfaces). Aspect ratio of hole diameter to length should be moderate; not more than 5:1 for producibility.

The outline should be as simple and rectangular as possible. Avoid radiuses or contours which can't be cut with a 0.062-in router bit. FPW layers should emerge from rigid areas perpendicularly to the outline, and the rigid edge at this point should be beaded or otherwise treated to protect the FPW from tear or cutting damage.

Specifications

Specifications and requirements for material and construction should be as open as possible. Do not narrowly define dielectric and adhesive choice and thickness (see Fig. 9-19 for an example of excessive specification); use performance specifications such as MIL-P-50884 for control of final product and dielectrics. Conductor foils should be as thin as possible consistent with electrical requirements; thicker foils require thicker adhesive layers which reduce thermal performance.

Tooling

Outlines in rigid areas may be toleranced closely because this contour be produced by NC routing with manufacturing and inspection coordinates oriented to nonplated tooling holes.

Unique tools required for RF are

- Slot routing program to predefine rigid areas at FPW exit points (see Chap. 9)
- Window dies to remove adhesive from flexible areas
- Filler dies
- Dies to cut FPW edges in flexing areas

FPW edges between rigid areas can be steel-rule die cut prior to multilayer stack-up; dies align to tooling holes in the panel borders (off-part coordinate origin).

Potting is seldom required in RF; forming or bending fixtures are also uncommon.

If bookbinder design is used, more tooling (vendor-supplied) is involved to accommodate the greater height of the in-process panels. These include special laminating and outlining fixtures and provision for greater Z dimension at electrical test and assembly.

Materials

The termination technique determines the material, which determines FPW cost. There are two classes of dielectric found in FPW: expensive systems which can withstand soldering, and lower-temperature, lower-cost systems which can't. If assembly will be by low-temperature mechanical means such as crimp, screw, compliant pin, or spring terminals, ZIF/LIF connectors, or conductive adhesive, dielectrics such as polyester and PEN are excellent choices. For survival in mass soldering, polyimides are best. (It is possible to solder to polyester-insulated FPW, but careful technique is essential, and yield will drop as a result of pad delamination.)

The intrinsic difference in film cost between polyimide and polyester, to take one example, is more than 10 to 1, but don't expect that to translate into correspondingly reduced FPW cost. Process labor, equipment, and overhead are the major cost elements in FPW built by the panel process: shifting from polyimide to polyester may yield only a 15 to 20% reduction in circuit cost.

Material is a more significant factor in roll-to-roll production economics. Furthermore, thermoplastics like vinyls and polyesters allow continuous roll production through base and coverlayer lamination cycles for significant process simplification. The combined effect is that, in roll-to-roll production, use of polyester and other thermoplastic dielectrics does significantly lower cost.

Because they're inexpensive, it's feasible to use polyesters in greater thicknesses—0.003 to 0.005 in, sometimes even 0.01 in—with PTF, which improves dimensional stability and reduces handling and tear problems. Polyester systems have low moisture absorption and relatively good dielectric properties (See Chap. 6); they are similar to polyimides except in cost and thermal resistance.

Since FPW material choice is interrelated with quality standards and assembly technique, calculation of overall cost benefit from material change is complex.

Effect of material choice on design

Material choice is defined in drawing notes which specify materials of construction and quality references. An important effect of materials on design is discussed in greater depth in the section "Dynamics," below. Other materials/design interactions:

- Dimensional stability of the dielectric through manufacturing processes is a critical consideration in FPW design.
- Larger pad areas with greater annular lands are beneficial if low-temperature dielectrics will be used with solder assembly because increased area provides more resistance to delamination and aids in dissipating heat. Larger pads may force more layers, which could nullify the slight material cost savings.
- High flex endurance requires thin foils and balanced construction (see "Dynamics" section).
- PTH terminated multilayers (and rigid-flex; see Chap. 9) are difficult to impossible to build in polyester materials and are best built in adhesiveless materials with low-expansion bond plies and adhesive layers.
- Artwork is largely material-unrelated; i.e., a good conductor pattern is a good pattern in any dielectric but compensation for etch factor will change if foil thickness changes. If artwork for metal-foil FPW is used for polymer thick film (PTF), important changes in conductor resistivity (increasing by approximately 30 times) and current-carrying capacity (reducing to essentially none) will result, and choice of termination technique will be limited to conductive adhesive and benign connector techniques such as ZIF/LIF.

Coverlayers, covercoats

FPW design includes choice of conductor insulation technique and design of openings or apertures. (See Chap. 8 for a detailed discussion of coverlayer and covercoat processes.) The lowest-cost insulation for a conductor pattern is a silk-screened covercoat; better aperture definition for high-density terminations can be achieved with photodefined coatings or films. The best dielectric and mechanical performance is provided by bonded dielectric films or coverlayers, but they cost more for materials and labor. The choice between covercoat and coverlayer depends on assembly technique, quality/reliability concerns, and environmental requirements.

Aperture design should take into account adhesive flow and alignment tolerances in the coverlayer technique and registration, photoresolution, and reduced crease resistance in the covercoat tech-

Coverlayer:	
Preparation	0.003 (drilling, die punch)
Registration	0.005 (pinning plus lamination slip)
Adhesive flow	0.005
	Total-0.013
Photodefined covercoat:	
Registration of artwork to pattern	0.005
Resolution	0.002
	Total-0.007

Figure 4-15 Aperture tolerances in inches.

nique (Fig. 4-15). Note that both techniques will misalign and reduce annular lands to the degree that the etched circuit distorts or shrinks.

Thicker conductor patterns (above 0.0014 in or "1 oz") are difficult to insulate with liquid covercoat systems, which tend to flow back away from the corners. Such patterns require use of photodefinable films or the coverlayer technique.

Conformal coatings can be applied over termination areas to provide protection against foreign material and accidental interconnections, but be aware that skill is required, cost will rise, and quality control is complicated by subjective evaluation of workmanship (see Fig. 4-13).

Dynamics

The FPW material choice and circuit design both affect flexural endurance.

Design guidelines

- Radius of flexure should be as large as possible with least angle of bend. Reasonable minimums: 24 times total thickness for single layer and 48 times total thickness for double-sided FPW.
- Where high flexural endurance is required, dielectric and foil layers should be as thin as possible and balanced in cross section, with conductors centered on the neutral axis of flexure between identical base and coverlayer insulations.
- Adhesive layers should be as thin and stiff (for foil-support) as possible; foil surfaces should be free of irregularities—pits, dents, etch faults, and scratches including pumice marks.

- Conductor runs should be routed perpendicularly to the axis of bending; foil grain, if known, should be perpendicular to the axis of bending.
- Bend lines should be located as far as possible from terminations or any mechanically secured or stiffened areas. Terminals or PTH should be at least 0.125 in outside the flex area.
- Wider runs in thinner foil are easier to manufacture, carry proportionately more current, and survive flexure better. Thinner foils allow thinner adhesives in the coverlayer, reducing insulation thickness.
- 0.001-in dielectric films are more than adequate for use below 500 V (except where chafing or abrasion are present; see Fig 4-16 for abrasion-resistant coating.)

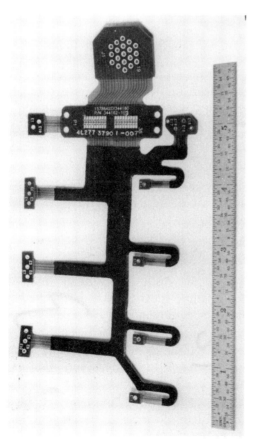

Figure 4-16 Simple rigid-flex circuit; single double-sided flex with pads-only hardboard caps. Black material is special abrasion protection, attached by pressure-sensitive adhesive. (*Courtesy V. F. Dahlgren.*)

- Flexing areas should have only one conductor layer. If there are two layers of conductors, runs on one side of the dielectric should be designed to lie in between runs on the other side (see Fig. 4-17) to minimize the offset from the neutral axis. If more than two layers are flexing, this region should be unbonded (see Chap. 9, "Rigid-Flex Manufacture"). If the PTH technique is used at terminal areas, it costs no more to also use it to jumper runs at the edges of the flex area so as to condense them all into a single layer (if possible).

Bending occurs in every FPW, if only at assembly. Therefore, it's a good idea to consider minimum bend radiuses (Fig. 4-18) and build in sufficient service loopage and locate mounting holes so that the FPW takes a gentle curve.

Materials considerations. Thinnest materials, strong, uniform bonds, and centered conductor layer with smooth, uniform surfaces provide longest flex life. Conductor surfaces are particularly critical—for optimum performance the surface should be free of scratches, dents, or other imperfections which will concentrate bending stresses, leading to early failure. Relatively stiff, cross-linked adhesives appear to improve flexural endurance compared to flexibilized systems, perhaps

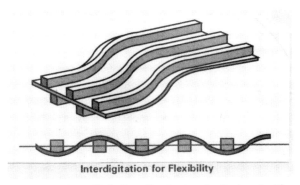

Figure 4-17 Interdigitation. Conductors in flexing double-sided designs should lie between each other (viewed in cross section) rather than directly above and below each other to avoid two-layer stiffness.

Single-sided	Double-sided	Multilayer
3 × thickness	10 × thickness	50 × thickness

Figure 4-18 Rules of thumb for minimum assembly bend radiuses in FPW.

because they more strongly keep the conductors centralized in the dielectric.

Qualification testing is important in dynamic applications because flex life is difficult to accurately predict from design and engineering data. Small differences in radius of flexure, material thicknesses, foil metallurgy, and surface smoothness can make major differences in lifetime. Test procedures must be carefully selected to assure they accurately represent the use conditions.

A standard flexural fatigue testing machine is the Model 2 FDF produced by Universal Manufacturing Company. This equipment, which applies reversing bends over an adjustable mandrel, will be found in any major testing laboratory and is required for MIL-P-50884 qualification.

An early test program* flexed sample FPW composed of 0.001-in polyimide base and coverlayer films and 0.0014-in copper runs to failure in a rolling U configuration 0.55 in in diameter. Failure could result from either fatigue or wear-out. Synopsis of results (percentages indicate portion of longest-life samples):

1. FPW built with in-house laminate: +8%
2. FPW flexed with coverlayer to the outside: +27%
3. Oxide treatment before coverlayer: +27%
4. Thicker coverlayer adhesive (tested coverlayer out): −23%
5. Thicker coverlayer to the outside (0.002 in versus 0.001 in): −75%
6. Acrylic adhesive type: +77%

Comments:

Item 1. Steinmetz attributed increased life of samples built with in-house laminate versus purchased laminate samples to uniformity of base and coverlayer adhesion when both are laminated by the FPW manufacturer. This does not seem to be a reasonable conclusion, since coverlayer was applied to untreated foil surface (in this test) while base lamination was to a treated surface.

Item 2. This is probably attributable to greater adhesion between conductors and base film (as compared to the coverlayer) as a result of a surface treatment on the base foil surface. Coverlayer adhesive was probably slightly thicker than the base adhesive (but not reported).

---

*James H. Steinmetz, "Comparison of Flexible Circuitry in a Dynamic Application," IPC TP 133.

Items 3 and 6. These improvements both appear to result from better bonding, therefore greater dielectric integrity.

Items 4 and 5. These results show the harmful effects of unbalanced construction. Note that flexure, in this test, was nonreversing, thus conductors in the thicker adhesive and coverlayer samples were continuously in compression.

Current, more stringent testing (as defined in IPC TM 650 method 2.4.3.1) consists of reversing flexure, usually around mandrel diameters which are chosen to produce failure within a standardized, short time/flexure span. Lifetimes of 100,000 cycles on a reversing $\frac{1}{4}$-in diameter mandrel are achievable in 0.001-in film/0.0014-in foil construction, with careful technique.

Forming

FPW can be designed for complicated bending and forming to create multilayered circuitry out of a single layer of FPW. Cutting and folding may be used to create circuitry which is several layers thick or nearly twice the panel length. Figure 4-19 shows a multilayer FPW assembly with soldered connectors; a single piece of FPW has been folded several times to create this dense harness.

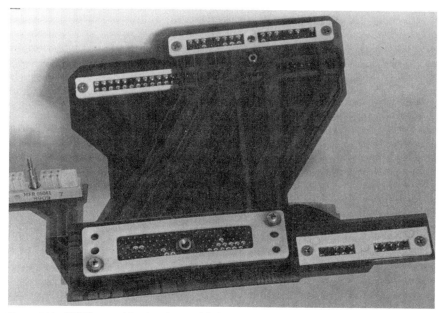

Figure 4-19 FPW assembly showing molded strain-relief rings.

It's not possible to precisely form FPW and expect it to stay as formed, because the materials are slightly elastic; roughly 15% springback can be expected in typical constructions.

Inspection to verify correct forming is both complicated and somewhat irrelevant, since the product is always reshapeable by hand. Tight bends are more durable, less manipulable, and therefore more drawing-definable.

Good practice, when forming is intended, is to leave as much conductor metal as possible at the bend location, because conductors are less springy than dielectrics. For the same reason, two-layer constructions are more formable and will spring back less than single-layer. The rule of thumb is this: more metal, less dielectric is more formable.

Provision for forming involves:

- Inserting alignment or "hash" marks or tooling holes (if tooling will be used) in the artwork
- Adding a cross-sectional view of the formed FPW to the drawings (but avoid tight dimensioning, unless the notation "shown in restrained condition" is used)

Stiffeners/strain relief

Where required for component support, for protection against shock and vibration, to help hold a preformed shape, or to aid sealing, areas of the FPW are strengthened by stiffeners. Figures 4-20 and 4-21 show a ring-shaped gasket section and a bar stiffener with mounting holes. Figure 4-22 shows multiple PTH termination areas in a rigid-flex circuit; each connector pattern has mounting holes in the rigid area which secure the FPW to protect the PTH area from strain.

Any insulation material can be used as a stiffener. Common choices are printed circuit laminates such as epoxy-glass, layers of polyimide film, or custom-molded pieces in polycarbonate or impact-resistant styrene. Alignment of stiffener to circuit is aided by hash marks designed into the FPW artwork and expressed as etched lines in the FPW, or by pin-alignment fixtures. Stiffeners can be bonded to the FPW by means of thermosetting adhesives during coverlayer lamination or by use of pressure-sensitive liquid or tape adhesives as an assembly process. For MIL-P-50884–certified products, the adhesive must be certifiable to IPC FC 233; if a prepreg is used, it must meet MIL-P-13949.

FPW has low mass and high tensile strength and hence is inherently resistant to shock and vibration. But where long unattached runs

Design 59

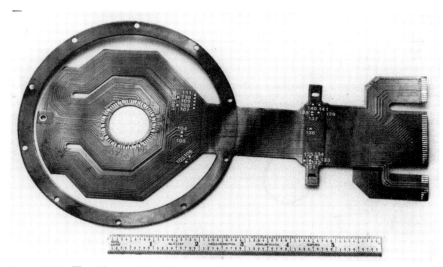

Figure 4-20 Flexible circuit with integral gasket. Two single-sided flex circuits are bonded together only in the ring stiffener and strain-relief areas. Termination is direct to bare pads. Circuits have 1-oz foil runs and Kapton dielectric. Center circle, 0.032 in in diameter is peck-drilled (note fuzziness). (*Courtesy V. F. Dahlgren.*)

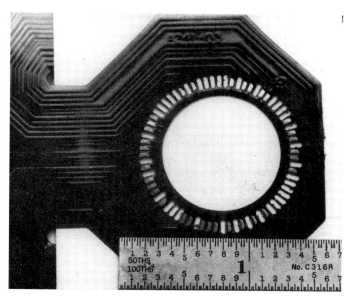

Figure 4-21 Detail of Fig. 4-20 shows clearance slots for access to area underneath circuit. The six holes in the ring were produced by precision jig. The tolerance on location is 0.001 in. Nomenclature identifies test point for troubleshooting. (*Courtesy V. F. Dahlgren.*)

Figure 4-22 Military rigid-flex motherboard for cannon-launched "smart shell" withstands extreme shock. Eleven hardboard areas are interconnected by multiple thicknesses of flexible circuitry. Board is manufactured in high volume.

are used or when the FPW attached to hardware items that may shift in such environments, support and strain relief are required. Figure 4-22 is a military harness which survives more than 10,000g during cannon lauch; note the stiffener/strain relief boards at each connection pattern, which serve to protect this interface from shock and mechanical stress.

Long runs of FPW are sometimes used as a handle or structural item during assembly operations (to carry subassemblies from station to station, for example) and considering stiffeners and strain relief is worthwhile for these situations as well.

Uninhibited access to throughholes in the FPW requires matching holes in the stiffener, which must be larger to allow for misalignment or adhesive intrusion.

Stiffener material, alignment, adhesive choice and bonding process are defined in drawing notes.

Panelization

Significantly improved high-volume production results when FPW circuits in panel form are bonded to a partially outlined panel of stiffeners in a process called *panelization*. The FPW/stiffener panel handles like a rigid board; can be fed through automatic component insertion and joining equipment and tested as a panel, then snapped apart into individual circuits. The panelization process:

1. A panel containing multiple partly outlined (or fully outlined and returned-to-panel) stiffeners is prepared.
2. Stiffener areas are coated with adhesive.
3. A matching panel of partially outlined FPW is bonded in register to the stiffeners.

Border areas of the stiffener panel provide the means for handling through automated component insertion, attachment, and test.

FPW with stiffeners can be thought of as a rigid-flex circuit; i.e., it has areas that are structural and areas that are flexible, joined by coextensive conductors as shown in Figs. 4-16, 4-22, and 4-23.

Stiffened areas of FPW may have PTH termination, in which case the stiffener and FPW are drilled for PTH as an assembly. For further discussion of combined rigid and flexible circuitry, see Chap. 9.

Edges of stiffeners should be radiused or beaded with compliant materials to protect FPW from abrasion and stress concentrations, as shown in Fig. 4-24.

Tear stops

FPW dielectrics must be dimensionally stable with high tensile modulus to protect conductors from mechanical damage. Materials of this type tend to have poor tear resistance (see Chap. 5, "Dielectric Materials," for properties). Unless precautions are taken, FPW outlines will have nicks, overcuts, and sharp corners which act as stress concentrators to initiate tearing. It's necessary for designers to take pains that outlines have radiused, not sharp, corners to minimize tearing and that slits end in relief holes. Product pictures in this book show radiuses at every corner, frequently enhanced by punched or drilled *tear-stop* holes; see Figs. 4-20, 4-21, 4-22, and 4-25 for examples of tear stops. Additional protection is provided by copper reinforcements. Mounting holes should be surrounded by copper foil, a no-cost addition in the conductor artwork.

Nomenclature

FPW artwork should include at least the part number and revision level of the circuit. This information is best expressed as etched or printed letters 0.1 in or more high, joined together and spaced far enough from electrified elements to avoid short circuiting. MIL-P-50884 expresses this as ". . . marking shall be produced by the same process used in producing the conductor pattern; or by the use of non-

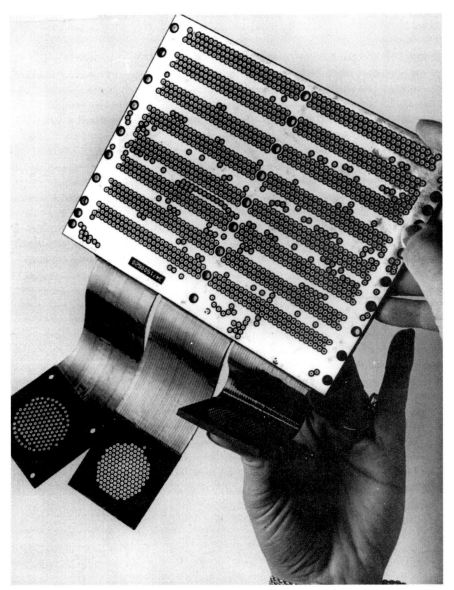

Figure 4-23 Military 14-layer rigid-flex board for avionics. Flexure of the head patterns provides 90° interconnect path and aborbs large dimensional tolerances. The main board surface has solder-coated ground planes; mounting holes have stress relief. (*Courtesy V. F. Dahlgren.*)

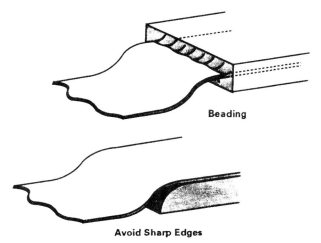

Figure 4-24 FPW must be protected from cutting or abrasion on sharp edges such as routed stiffeners or rigid areas of RF circuitry. This can be accomplished by applying a semirigid fillet, or bead, of potting material or by chamfering or radiusing the sharp corner.

conductive, permanent fungistatic ink or paint applied to the board or to a label which is applied to the board."

It's good practice to identify key pins at connector patterns and any location where ambiguity could lead to incorrect assembly. If space can't be found within the conductor artwork, then separate nomenclature artwork (with characters large enough for legibility) for screen print or ink stamp will be required, with a drawing note defining location, ink type, and color. External nomenclature is less desirable because of added labor and possibility of error, but is needed for date codes and similar changeable information.

Added pads

Thought should be given to the use of extra termination points to facilitate test or circuit change. Added test points on standard grid dimensions can be used to assist in electrical test of unusual pattern layouts. Where potted connectors are used or anywhere that the assembled FPW termination point is hard to contact for test purpos-

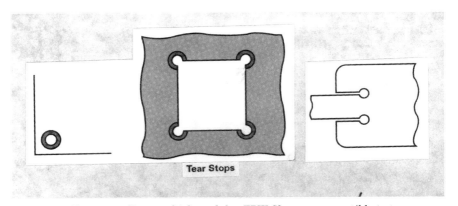

Figure 4-25 Tear stops. Because high-modulus FPW films are susceptible to tear propagation, inside corners should end in smoothly radiused holes or other stress-dissipation features. Inclusion of etched copper reinforces this protection.

es, an added test pad (with suitable identifier) is desirable. If change in the electrical interconnect pattern is anticipated, extra termination points may be designed into the pattern to be jumpered as desired. If frequent component rework/replacement, which will eventually destroy FPW pads, is anticipated, a secondary set of terminations can be provided as future replacements.

Coupons. Coupons are invaluable for control of manufacturing process, documentation of product quality, and development of engineering data bases. A variety of standardized designs are available from the IPC, military standards, and other sources, and should be incorporated into production composites to take advantage of otherwise wasted panel area. Examples of coupon types:

1. Flexural endurance test specimen
2. PTH integrity by cross-sectional analysis:
    a. Etchback
    b. Material thickness: plating, foil, dielectric
3. Coverlayer/covercoat adhesion
4. Dielectric properties:
    a. Insulation resistance
    b. Dielectric strength
    c. Dielectric constant
5. Solderability
6. Thermal stress:
    a. Solder dip
    b. Hot oil

Tolerances and multilayer design

Alignment is a constant FPW design and manufacturing concern. It's required for conductor imaging, when coverlayers and covercoats are applied to conductor patterns, and for multilayer constructions. Tooling holes and aligning pins are the traditional technique for assuring adequate alignment or registration between tool and pattern (see Chap. 8); alignment of artwork to produce multiple conductor layers which must align or register for multilayer assemblies follows the same practices. Because FPW materials are inherently unstable to some degree, perfect multilayer alignment cannot be achieved.

Designs must accommodate these variations:

- After-etch shrinkage: 0.001 in/in
- Artwork variation: 0.0002 in/in from humidity and temperature effects (plotting error plus film base instability, unless the base is glass)
- Pinning: each occurrence, 0.002 in
- Lamination: 0.003 in (up to 24-in panel, with good technique)
- Hole-to-hole in drilled patterns: 0.002 in

In the worst case that all variations can sum to, it's clear that only very simple designs with large overlap allowances can be built. Artwork factoring to compensate for material movement after etch is almost mandatory; thoughtful material choice at the design stage, plus clear definition of material and quality requirements in drawing notes, is essential to close the control loop. See "Dimensional Stability" in Chap. 3 for further discussion.

Optical alignment to etched fiduciaries (see Figs. 4-26 and 4-27) is a method for reducing alignment errors which reduces the effect of artwork pinning errors and after-etch shrinkage.

Design aids for alignment

A basic design strategy is use of common tooling holes or fiduciaries which are carried in the data base, inserted into each circuit layer, coverlayer or covercoat design and included in product manufacturing instructions and dimensional drawings. In double-sided or multilayer designs, each layer of artwork includes common tooling holes which, when aligned, produce correctly registered conductors. In the PTF technique, multiple screen patterns are sequentially printed (the previous image is dried or cured between imprints) onto the substrate, aligned by means of retracting tooling pins in the vacuum platen.

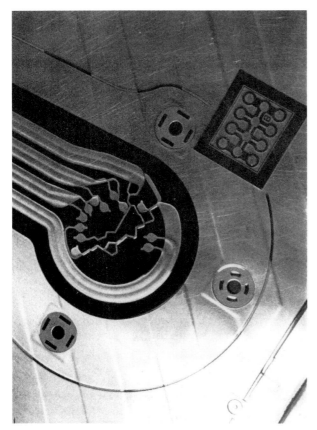

Figure 4-26 Close-up of cluster registration. Note three optical fiduciary patterns which have been automatically centered and punched; note also shielded circuitry and ventilation pattern visible in the embossed foil. Rectangular pattern with the nomenclature B is a test coupon. (*Courtesy Teledyne Electronic Technologies.*)

Composites

An arrangement of multiple repeats of a circuit design within a production panel for efficient production is called a *composite*. Conflicting forces work here: the lowest circuit cost, assuming a high yield, results from the maximum number of circuits per panel; the highest yield results from good dimensional stability, which means fewer circuits and more stabilizing copper.

Figure 4-27 Ushio optical punch with robotic guidance. (*Courtesy Flex Technology Inc.*)

Area/cost

FPW is manufactured in panels which range from 9 × 12 in to 18 × 24 in, with usable size reduced by process borders and allowance for tooling holes and process control coupons as required. Labor is relatively independent of panel size; material cost depends on area. Cost of FPW is panel cost divided by number of circuits per panel. If a panel sells for $100, and there are 10 copies of the design in the composite, per-circuit cost is $10.00. Figure 9-17 shows a well-composited, two-up RF circuit with coupons.

This is a basic rule of thumb for FPW design: area = cost. Area in this context means *number of circuits per composite*. Nesting, efficiently fitting circuits together so as to yield the greatest number of circuits within the defined panel size, is the controlling concept. It's not necessary to laboriously planimeter circuit area to figure how many can be packed onto a panel: simply make copies of the paper dolls and see how many fit into the panel dimensions. Note that adding area to a design doesn't add cost until it reduces the number of circuits that fit on a panel; up to this point, design changes are no-cost.

Important points to remember:

- Projecting arms and irregular shapes cost more if they reduce the number "up" (circuits repeated) on a panel.
- Circuit size and cost are quantum-defined by panel dimensions.

Just how tightly to pack circuit patterns within the borders, and whether to lay them in both across and parallel with the machine direction of the material, is a difficult decision. Sufficient space must be left between circuits for outlining, PEP (optical tooling targets), process control coupons and nomenclature. Sometimes interaction between circuit layout and coverlayer design suggests a more efficient composite—for example, if the coverlayer includes a rectangular cutout at one end of the circuit, aligning the circuits so that all those ends lie in a straight line allows use of a sheet of coverlayer material, rather than complex cutouts.

Remember that coverlayers may be in composite form, too—a panel of coverlayers designed to align with a panel of etched circuits—or may be in individual pieces. Individual coverlayers allow better alignment between coverlayer and circuit because each circuit is individually aligned, and improve utilization of coverlayer material. More labor is required, of course, both to prepare individual coverlayers and to align them.

Circuit artwork should include hash marks for alignment of coverlayers as well as for forming or bending tools or stiffeners.

Borders

A good rule of thumb for 18-×24-in panels is a 1-in border of stabilizing copper. This reduces the usable area by 19%, but improves stability significantly.

Solid borders interfere with venting of air and volatiles during coverlayer lamination; for this reason, in the PWB business, borders are broken apart by vent channels. Glass-reinforced PWB materials are about twice as stable as FPW materials and therefore benefit less from border reinforcement; in FPW materials, use of broken or vented borders results in greater variation from artwork to part and a wider range of variation. Solid borders used with vacuum-assisted lamination (see Chap. 8, "Manufacturing Processes") are strongly suggested.

In one study based on a typical moderate-density double-sided layer for a rigid-flex board, shrinkage and variation of shrinkage from artwork to etched, coverlayered 0.0014-in foil circuit looked like this:

	Solid border, in		Vented border, in	
Artwork size, in	Average	Range	Average	Range
6.75	6.747	0.007	6.745	0.009
10.503	10.496	0.011	10.496	0.013

The solid border parts shrank an average of 0.003 in in the 18-in direction of the panel and 0.007 in in the 24-in direction, with maximum variations of 0.007 and 0.011 in; the vented border parts shrank an average of 0.005 in and 0.007 in, with ranges of 0.009 and 0.013 in. Variations are a result of residual laminate stress (i.e., latent defects in the purchased material) plus stresses from handling and the coverlayer process.

Stability increases with greater conductor density and thicker copper and decreases with fewer conductors and thinner copper. In this study, 0.0028-in and 0.0014-in layers shrank as follows:

Artwork nominal size, in	Average measured artwork size, in	
	0.0028-in layers	0.0014-in layers
6.75	6.748	6.748
10.503	10.501	10.498

Be aware that shrinkage is very design-specific; i.e., where there's a good copper density in a given direction, shrinkage will be reduced but will probably increase across the conductor bundle because of greater coverlayer stress caused by conforming around the etched pattern.

Factoring

Examination of shrinkage data leads to the conclusion that artwork *factoring*—expanding it to compensate for expected shrinkage—is a wise idea. A suggested starting point is a scale factor of 1.0005, i.e., an increase of 0.0005 in per inch. In the shrinkage study cited, this would raise the averages across the 6-in dimension by 0.003 in and across the 10-in dimension by 0.005 in, nicely centering the results on the artwork nominals for the solid border case. Factoring won't correct for variations, though—a conservative design with adequate annular rings and as much production tolerance as possible is needed for this.

Shielding

Varying current flow produces electric fields which broadcast energy. Protecting against intrusion of such energy onto nearby conductors is the purpose of shielding. The need for shielding increases with:

- Increasing frequency and magnitude of current flow
- Proximity between disturbing and sensitive conductors
- Increasing impedance
- Decreasing signal level
- Increasing amplification

The consistent, production-reproducible nature of FPW conductors legitimizes empirical design development. It's practical and cost-effective to build an unshielded circuit and test it, because more often than not, the simplest design will meet requirements, and if a test circuit passes, production FPW will also pass.

Shielded FPW is particularly cost-effective as a wire harness replacement because shield planes in FPW can be designed and artwork-controlled to protect one or any number of runs, and multiple independent shields can be etched from a single foil sheet, each segment individually grounded. Electrical capacity between shield and conductor is controlled by design parameters of dielectric choice and thickness.

Internal-external disturbances

Disturbing fields may be generated by another circuit in the FPW design (internal disturbances or crosstalk) or may originate from external sources.

Internal. If the disturbance comes from another FPW circuit, the best strategy is to first isolate disturbing from sensitive conductors—locate these groups as far from each other as possible, into different pieces or layers in multilayer designs—and test to see if this provides sufficient isolation. If it doesn't, the next step up is to interpose grounded conductors between sensitive and disturbing conductors. If this fails, application of shield planes surrounding the sensitive runs is necessary.

External. Shields are required to protect against external disturbances. The principles of shielding are quite complex: shield conductivity and low-resistance attachment to circuit ground are important, but how shields interact with an external disturbing field is difficult to express in terms useful for the designer. Effectiveness is equally difficult to quantify and therefore requires testing. Unfortunately there's no standard test, and circuit operating conditions can range across a wide frequency and power range, as can questions of terminating resistance and separation between source and test conductor. The subject can only be generally discussed with the intention of setting forth guidelines.

Chapter 6, "Conductive Materials," lists common shield materials; these include solid and patterned metal foils, PTF overprints and screen layers. Shield conductivity is important, and so is thickness, because disturbing fields penetrate through shield layers to a depth—so-called skin depth—which depends on frequency.

Shield configurations. Shield planes are layers or sheets of conductive material which are connected to a reference circuit ground. In order of increasing cost, efficiency, and stiffening effect, shield strategies are

Loose (unbonded)

Single-sided, bonded

Double-sided, bonded

Loose shields are layers of conductive material, usually but not necessarily insulated, that are inserted in a stack of FPW layers or sandwiched around a sensitive layer and loosely attached by taping or bundling. Single-sided but bonded shields are integrally joined to one side of the FPW; the circuit will be stiffened and flexural endurance reduced because conductor runs are no longer located at the neutral axis. Double-sided, bonded shields are more effective but stiffen the FPW to the point where it should be considered formable rather than flexible.

Because electrical capacity between shield and conductor is important, loose shields have variable effectiveness. Single-sided, bonded shields are stable, but protect against disturbing fields from only one direction; where the situation assures that disturbances come from only one side, single shielding is a good choice. The best method is double shielding, which offers maximum protection but at considerably increased cost and with serious effect on flex life.

Another shielding technique is to wrap the FPW in conductive tape, which self-adheres and can be formed into a complete cage which includes circuit edges. This method requires considerable skill to produce a neat envelope but is very effective.

Adhesion between shield and over- and underlying layers is improved if the shield plane is etched or apertured to allow the adhesive to penetrate through and directly bond over- and underlying layers. Aperturing improves flexibility but degrades efficiency, as shown in Fig. 4-30.

Low-resistance termination of shield to ground is important; high-resistance paths degrade effectiveness. Ground connection rules are

1. No loops—straight from shield to ground.
2. Keep as short as possible.
3. Single interconnection is correct for low frequency (less than 1 MHz); multiple is better for higher frequencies.

Figure 4-28 shows a double-shielded cable which uses PTH interconnections spaced 1 in apart in three rows to attach shield planes to a ground conductor in the signal (central) layer. Direct interconnection—for example, through an etched extension from a shield which intrudes into a connector pattern to be directly soldered to the ground pin—is a straightforward and effective grounding technique.

PTF shielding. PTF shield layers can be screen-printed directly onto the FPW or onto a coverlayer which is later laminated onto the FPW. If PTF is applied to the adhesive side of the coverlayer, the advantage is that the coverlayer acts as an abrasion protector for the PTF, but the method requires extra steps to interconnect shield to ground. Interconnection of direct-printed PTF to circuit ground can be by "wet overprint" onto an etched ground terminal. Figure 4-12 shows a series of such terminals in this in-process circuit.

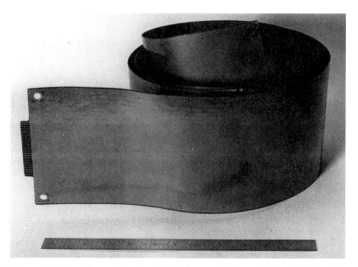

Figure 4-28 Sixty-conductor, 48-in-long, 50-Ω stripline FPW with vacuumtight construction is used in test instrumentation. Tabs visible at the ends adapt the FPW for ZIP termination. (*Courtesy Teledyne Electronic Technologies.*)

Comparisons

Figures 4-29, 4-30, and 4-31 present the results of tests of various shielding methods.

Conclusions from the simple study in Fig. 4-29 are

- Grounding adjacent runs provides some degree of protection
- Solid foil shields are very effective in reducing electrostatic pickup
- Aperturing reduces shield effectiveness

The results in Fig. 4-30 show that lower resistance and greater thickness of the braided wire harness shield provided better protection at low frequencies. Above 200 MHz, the PTF shield, in spite of higher resistance and less thickness, provided better protection, probably because it's a continuous coating without apertures. Erratic variations in the results are a consequence of resonances.

The data in Fig. 4-31 are somewhat corrupted by resonances and random placement of wires in the conventional harness. Above roughly 200 kHz the shielded FPW cable is comparable or superior to the shielded wire harness and rapidly improving, compared to unshielded FPW, at frequencies above 100 MHz.

Controlled Impedance

Precisely controlled, uniform spacing between shield and conductors and careful selection of dielectric material is required in controlled-impedance circuitry. In this class of FPW the electrical capacity per

---

Test method: A 10-in length of FPW was placed in a uniform electric field produced by two electrodes 5 in square, spaced $9/16$ in apart and energized by a 1-kHz source. Voltage pickup on the test conductor was filtered to eliminate spurious signals and measured by a high-impedance instrument.

Shielding technique	Measurement, dB
Unshielded conductor voltage pickup	0 (reference)
With one adjacent conductor grounded	-2.7
With both adjacent runs grounded	-9.8
With one apertured shield (0.0014-in copper)	-20
With two apertured shields (0.0014-in copper)	-26.3
With one solid shield (0.0014-in copper)	-33.7
With both sides shielded (0.0014-in copper)	-55
Reference: RG-59U coaxial cable	-50

Figure 4-29 A practical study of the effectiveness of shield techniques in reducing externally induced voltages in FPW.

	Effectiveness relative to unshielded wire, dB		
Frequency	Unshielded FPW	Shielded FPW	Shielded wire
10 kHz	1	64	77
50 kHz	1	22	36
100 kHz	2	20	39
200 kHz	1	20	28
500 kHz	2	16	33
1 MHz	1	24	33
3 MHz	1	3	30
5 MHz	5	9	14
10 MHz	2	24	46
20 MHz	0	29	36
60 MHz	−3	21	38
100 MHz	−4	28	34
200 MHz	−8	30	14
500 MHz	0	20	13
1 GHz	14	31	16
1.5 GHz	20	38	18
5 GHz	6	30	10
10 GHz	2	6	3

Description: Four samples—conventional wire harness with and without braided shield and flat conductor cable with and without silver conductive (PTF) ink shield. Samples were exposed to disturbing external fields from 10 kHz to 10 GHz in a screen room. Terminating resistors were not used. All data was normalized to the unshielded wire harness.

Figure 4-30 Test of PTF shield effectiveness.

unit length is uniform and matched to the effective inductance of the conductor runs.

Impedance determines transmission efficiency: when the impedance of the interconnecting circuit matches the impedance of the devices interconnected, signal transfer is best and reflections are minimized. When the impedance is mismatched, part of the energy is reflected; the reflection percentage is given by the ratio of impedances:

$$R = \frac{Z - Z_d}{Z + Z_d}$$

where Z is the FPW impedance and Z_d is the impedance of the device. Thus, if we have a 50-Ω device and it's connected to a 40-Ω FPW run, 11% of the energy (10/90) will reflect and 89% will be transferred.

Description of test setup:
Sample length: 6 ft
Termination resistance: 50 Ω
Interposed grounded runs
Shield techniques: for FPW, PTF; for wire harness, braid

Frequency	Isolation, dB			
	Shielded		Unshielded	
	FPW	Wire	FPW	Wire
10 kHz	60	68	60	68
50 kHz	59	62	59	61
100 kHz	60	59	58	57
200 kHz	57	54	59	52
500 kHz	49	47	56	44
1 MHz	39	34	53	31
3 MHz	30	33	44	27
5 MHz	26	31	40	24
10 MHz	23	29	36	20
20 MHz	25	18	32	18
60 MHz	28	21	38	12
100 MHz	39	13	14	9
200 MHz	49	12	17	25

Figure 4-31 Internal disturbances; isolation between adjacent, terminated conductors in FPW and conventional wire harnesses, shielded and unshielded.

Designs

FPW impedance designs are shown in Fig. 4-32.

Coplanar has no shield planes—an adjacent conductor run serves that purpose. This is the most flexible configuration because it's single-layer, is thinnest, and has good dynamic performance because the conductors can be centered in a balanced construction. It is relatively susceptible to external interference and has high crosstalk, i.e., unwanted transfer of signal to adjacent conductors, because of shared ground currents. Crosstalk can be reduced by use of two grounded runs for each signal run, but FPW width will increase by 50%.

Microstrip, with one shielded side, is better protected from external effects but more expensive to build and stiffer; it isn't good for dynamic use. Crosstalk is less than with coplanar design.

The best design for stable transmission in hostile environments is the double-shielded *stripline* configuration. This is the most expensive to build, requires greatest dielectric thickness and thus is stiffest (don't use in dynamic applications), but has excellent performance.

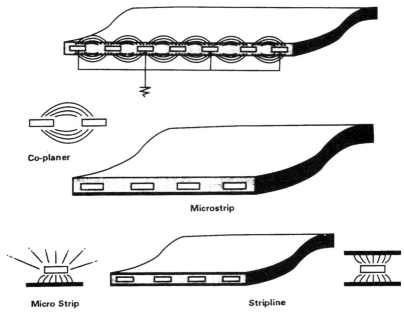

Figure 4-32 The three FPW controlled-impedance designs are *coplanar* (all circuitry in one plane), *microstrip* (a single shield plane), and *stripline* (both sides shielded). In coplanar design, alternate conductors are grounded. The electric field is loosely contained in the vicinity of the FPW. Microstrip has good field containment on its shielded side and all conductors can be signal carriers. Stripline has the best field containment and best isolation from external disturbances because all field lines lie within the shield planes.

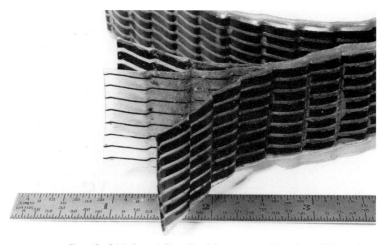

Figure 4-33 Detail of 93-Ω stripline flexible circuit cable. Overall length is 10 in. Center conductors are 0.008-×0.025-in copper in polyethylene dielectric manufactured by roll lamination. Shields are 1-oz copper in polyethylene manufactured by the print-and-etch technique. The spacing dielectric is urethane foam for low dielectric constant. Periodic quilting provides controlled distortion to allow severe flexing without discontinuity.

Each conductor is well-isolated by the shield planes, therefore crosstalk is minimized.

See Figs. 4-28 and 4-33 for examples of FPW stripline transmission cables. Figure 4-33 shows details of eight 93-Ω conductors in a cable 10 ft long; Fig. 4-28 is a 48-in-long stripline circuit with 60 conductors at 50-Ω impedance.

Formulas for calculating impedance are approximate at best and have boundary conditions beyond which they're not accurate; they're usable only within certain ratios of conductor and dielectric dimensions. Development of precise designs requires prototype fabrication and test, followed by adjustment to conductor width or dielectric thickness if required. The following spreadsheet presentations are based on formulas which are 90% or better accurate and adequate for preliminary design work. They're included to show the differences between coplanar, microstrip, and stripline approaches to a 50-Ω controlled-impedance FPW design, and the sensitivity of these techniques to conductor width variations.

Spreadsheets

Coplanar Configuration

Conductor		Dielectric		Impedance, Ω, coverlayered coplanar
Width w, in	Spacing s, in	Thickness t, in	Constant K	
0.0040	0.0080	0.0014	3.7000	95.934
0.0050	0.0080	0.0014	3.7000	85.335
0.0060	0.0080	0.0014	3.7000	76.278
0.0070	0.0080	0.0014	3.7000	68.370
0.0080	0.0080	0.0014	3.7000	61.353
0.0090	0.0080	0.0014	3.7000	55.047
0.0100	0.0080	0.0014	3.7000	49.319
0.0110	0.0080	0.0014	3.7000	44.074
0.0120	0.0080	0.0014	3.7000	39.235
0.0130	0.0080	0.0014	3.7000	34.745
0.0140	0.0080	0.0014	3.7000	30.557
0.0150	0.0080	0.0014	3.7000	26.632

General formula:

$$\text{Impedance} = \frac{C}{\sqrt{K}} \ln\left(\frac{3.1416s}{(w+t)}\right)$$

where C ranges from 110 to 130.

Microstrip Configuration

Conductor		Shield		Impedance, Ω,	
Width w, in	Thickness t, in	Separation s, in	K	Surface	Coverlayered
0.0050	0.0014	0.0040	3.7000	57.2801	61.13
0.0060	0.0014	0.0040	3.7000	51.9632	55.45
0.0070	0.0014	0.0040	3.7000	47.2924	50.46
0.0080	0.0014	0.0040	3.7000	43.1277	46.01
0.0090	0.0014	0.0040	3.7000	39.3699	42.00
0.0100	0.0014	0.0040	3.7000	35.9466	38.34
0.0110	0.0014	0.0040	3.7000	32.8031	34.98
0.0120	0.0014	0.0040	3.7000	29.8971	31.88
0.0130	0.0014	0.0040	3.7000	27.1952	28.99

Surface:

$$\frac{87 \ln\left[\dfrac{5.98s}{0.8w + t}\right]}{\sqrt{K} + 1.41}$$

Coverlayered: $\left(\dfrac{1}{\sqrt{k}}\right) \bullet 54.4 + 98.8/1000 \bullet \text{separation} \bullet \ln (5.97 \bullet \text{separation}) \, (\bullet 8 \bullet \text{width} + \text{thickness})$

Surface means the topside of the conductors is exposed to air. Coverlayered impedance is lower because replacing the air dielectric which is on the topside of a surface circuit with a coverlayer for which $K = 3.7$ raises capacity, thus lowers impedance.

(From M. Saubert and D. Snyder, "Packaging Engineers Face Conflicting Demands," *Electronic Packaging and Production,* June 1985, pp. 152–153.)

50-Ω design examples

There is an infinitude of possible conductor sizes and dielectric choices for any configuration and impedance. To illustrate the relationship between conductor size, dielectric thickness, and impedance, let's assume we're designing 50-Ω transmission lines in FPW, and wish to keep the conductor width to around 0.005 in because we don't want to get involved in very fine line production. In a coplanar configuration the dielectric thickness isn't a primary factor, and conductors 0.01 in wide, 0.0014 in thick, spaced 0.008 in away from each other yield about 50 Ω. With base and coverlayer dielectrics, total overall thickness is about 0.006 in. In microstrip, embedded version (i.e., with coverlayer), 50 Ω results from conductors 0.007 in wide in 0.0014-in foil spaced 0.004 in from the shield; overall thickness with 0.0014-in shield foil, base dielectric, and coverlayers is about 0.010 in.

Stripline Configuration

Conductor		Shield		Capacity, pf/ft	Impedance Z, Ω
Width w, in	Thickness t, in	Separation s, in	K		
0.0030	0.0014	0.0150	3.7500	4.2483	62.44
0.0035	0.0014	0.0150	3.7500	4.9563	59.33
0.0040	0.0014	0.0150	3.7500	5.6643	56.52
0.0045	0.0014	0.0150	3.7500	6.3724	53.93
0.0050	0.0014	0.0150	3.7500	7.0804	51.55
0.0055	0.0014	0.0150	3.7500	7.7885	49.33
0.0060	0.0014	0.0150	3.7500	8.4965	47.27
0.0065	0.0014	0.0150	3.7500	9.2045	45.33
0.0070	0.0014	0.0150	3.7500	9.9126	43.51
0.0075	0.0014	0.0150	3.7500	10.6206	41.79
0.0080	0.0014	0.0150	3.7500	11.3287	40.15
0.0085	0.0014	0.0150	3.7500	12.0367	38.60
0.0090	0.0014	0.0150	3.7500	12.7448	37.13
0.0095	0.0014	0.0150	3.7500	13.4528	35.72

$$\text{Capacity} = 5.4K \left(\frac{w}{K}\right)\left(\frac{w}{s - t/2}\right)$$

$$\text{Impedance} = \frac{60}{\sqrt{K}} \ln\left[\frac{(0.67)(3.1416w)}{w/(s - t/2)}\right]$$

(From M. Saubert and D. Snyder, "Packaging Engineers Face Conflicting Demands," *Electronic Packaging and Production*, June 1985, pp. 152–153.)

Stripline requires much more dielectric thickness: a 0.0055-in wide conductor in 0.0014-in foil requires 0.015 in of dielectric between the shield planes. Overall thickness, including 0.0014-in shield and coverlayer, is around 0.021 in.

Relative stiffness is directly related to thickness, factored by the number of copper layers. Coplanar design is indistinguishable from single-layer FPW; microstrip will be noticeably stiffer; stripline will behave more like a copper foil strip—i.e., it will be formable rather than flexible.

Summary:

	Coplanar	Microstrip	Stripline
Conductor width, in	0.01	0.007	0.0055
Overall thickness, in	0.006	0.01	0.021
Conductor layers	1	2	3
Stiffness	Low	Moderate	High

Other design considerations in controlled impedance are capacity and velocity of propagation (functions of dielectric choice) and phase linearity, which means all conductor runs of equal length.

Transitions

Be aware that correct transition to an FPW controlled-impedance conductor from connectors or components is at least as important to performance as conductor impedance. More often than not the mismatch through a poorly chosen connector or badly designed component attachment creates more signal loss and larger reflections than would occur from the worst possible conductor/shield/dielectric FPW design. Design of test fixtures for controlled-impedance FPW is a complex task for this reason. Time-domain reflectrometry (TDR) is used as an attempt to avoid transition problems and produces test results that are reassuringly crisp and definitive but can be very misleading.

Design of good launches onto FPW runs is well beyond the scope of this book and is specific to the hardware. The general idea is to maintain desired impedance as the cross section changes, for example, from planar FPW to coaxial connector, by tapering conductor runs and dielectric thicknesses. Smoothness of taper defines the upper operating frequency: to get into the gigahertz range (now appearing in PCs) requires very precise, elegant transition. There are connector and component designs available for interconnection with printed circuitry. For further details of the interface design, contact the manufacturers.

Documentation

In the beginning, artwork was created from master patterns carefully hand-inked onto huge panels of stable material. Drawn 4 to 10 times oversize by skilled hands able to locate within 0.007 in, then photoreduced to working size, these layouts produced very accurate artmasters. Today, virtually all FPW design is carried out on CAD/CAM workstations with powerful software routines and magnetic data transfer. The data file is a controlling document of the design.

Artwork

Output from the CAD system is fed into a photoplotter to produce master artwork which is in turn contact-printed to make working copies. When greater accuracy is needed, so-called first-generation artwork—direct from the plotter—is used. Accuracy of replicating the database into the film master depends on the plotter, film base, and

Plotter	Property	Plastic (Estar*)	Glass (soda-lime)
0.0002 in	Temperature coefficient	18×10^{-6} per °C	8.1×10^{-6} per °C
	Humidity coefficient	27×10^{-6} per % relative humidity	No effect

*Registered trademark of Eastman Kodak Company.

Figure 4-34 Artwork tolerances.

environmental controls. Plastic film bases are sensitive to both temperature and humidity change (Fig. 4-34). Glass is most stable but raises cost and complicates handling and alignment.

For example, if the temperature difference from plot to use is 5°C, plastic-based artwork will change by 90 ppm or 0.9 mils in 10 in; glass by 40.5 ppm or 0.4 mils in 10 in. If relative humidity changes by 10%, plastic change is 270 ppm or 2.7 mils in 10 in.

Drawings

Production of FPW requires drawings which set forth dimensions, inspection requirements, and materials of construction. Where required, cross sectional views are added for clarity. Other drawing-defined items: unique quality provisions and assembly processes such as potting, forming, beading, and nomenclature.

More often than not, FPW drawings are far too detailed and specific, perhaps from designer ignorance of production methods and fear of underspecification. It's good policy to apply close tolerances *only* within hole clusters that must fit hardware pin clusters—connectors, relay sockets, etc. Hole patterns will be produced by NC drilling or tooling, and good precision is needed, possible, and inspectable. Overall lengths can be governed by minimums, or fractional tolerances, because FPW between termination patterns should be slightly too long—in these areas, precision isn't needed. Remember that it's not easy to measure over longer distances on FPW without laborious fixturization, simply because the product is flexible and therefore doesn't lie in a single plane. Figures 4-20 and 4-21 illustrate gross overspecification—the tolerance on the location of the clearance holes in the gasket ring is 0.001 in, forcing use of hardened tooling both to create the holes and to inspect their location. As shown in the picture, the circuit naturally bends and distorts, negating the value or relevance of precision hole location.

Generic quality standards (see Chap. 10, "Standards and Specifications") are invoked by reference.

A final schematic should be prepared from the database used for artwork preparation and submitted as part of the data package. This interconnection playback should identify any unintended changes for early correction.

For similar reasons, the drawing package may include a bill of material which includes FPW and assembly items.

Double dimensioning. A common problem in FPW production is dimensional conflict between artwork and drawings. Artwork is a production tool which directly (except for etch factor and reproduction errors) controls circuit size. The designer is urged to carefully overlay artwork and drawing requirements to assure that no conflict exists.

Drill programs

Location of drilled holes to a reference datum is provided in magnetic data form or can be directly fed to NC drilling equipment in shops which are so equipped. Custom dies may also be required and are defined by drawings with notes and tolerances.

Test data

Electrical test based on the designed interconnect pattern provides double-check assurance that the design is correct. Direct and magnetic data transfer are both used. Optical inspection of etched details is facilitated by direct feed from CAD data to the inspection station. Alternatively, design rules or a "golden standard" reference part may be used, but as with electrical test, setup and inspection are greatly enhanced through reference to CAD data.

Outline data

Steel rule dies (see Chap. 8, "Manufacturing Processes") are widely used to outline FPW circuitry, stiffeners, adhesive layers, coverlayers, and so forth. These tools are relatively inexpensive and adequately precise. See Fig. 4-2 for standard outline tolerances. CAD data can be supplied to vendors of steel rule dies either in floppy disk form or by modem. So-called machined steel rule dies operate on the same knife-edge cutting principal, but are produced by machining from a block of steel rather than by inserting knife-edged bands of steel into a carrier and thus are more accurate—and costly. For best precision and life, matched dies may be used; such tools reduce cost in high volumes when operated in automated punching equipment.

Summary

FPW design involves interacting mechanical, electrical, and materials issues. It is similar to printed wiring board design, complicated by the third dimension. Considerable up-front engineering and design effort plus procurement of expensive tooling is required before production of FPW can begin, factors which interfere with wider use.

The termination technique—typically solder—is enormously important in FPW design. The chosen method determines the materials of construction, number of conductors per layer, and circuit cost.

FPW is particularly cost-effective as a replacement for shielded or controlled-impedance wiring. Its use brings significant reduction in installation labor while improving performance and consistency from assembly to assembly.

Good design takes the inherent instability and handling problems associated with thin FPW materials into account. Lowest circuit cost comes from efficient composite layouts, thoughtful circuit design to achieve high conductor density, and reasonable quality requirements.

Multilayer construction provides the high interconnectivity needed in dense packages but requires added attention to important design concerns of layer-to-layer registration, PTH processing, and thermal stress survivability. Rigid-flex design—the most complex of FPW variations—brings more concerns from processing mixed rigid and flexible materials, more complicated outlining, more tooling for slots, fillers, and venting, plus maximum inspection and quality surveillance.

Chapter 5

Dielectric Materials

Introduction

Flexible printed wiring (FPW) is a stratified product. It consists of layers of insulation, adhesive, and conductive materials put together in a sequence as required by a particular design. This chapter deals with the dielectrics—the films and adhesives used in *base* and *coverlayers,* as well as *bondplies* and *cast adhesives*—with a few words about potting compounds and conformal coatings. Conductive materials are discussed in Chap. 6.

In FPW, as in printed wiring boards (PWBs), the dielectric serves both electrical and mechanical purposes; the designer chooses a system of film and adhesive to get the combination of properties which is correct for the application. As is typical in mature industries, the FPW business has accumulated a cluster of language and customs pertaining to dielectric choice which are confusing to the newcomer. In this chapter we'll try to set the materials issue straight.

Elements

Conventional FPW consists of five distinct layers of material—a base dielectric film, conductor layer, a coverlayer film, and two adhesive layers. One of the adhesive layers bonds conductors and base dielectric together to form a base laminate; the second one joins coverlayer dielectric to the etched pattern. It's good practice to have matching properties in base and coverlayers; for this reason FPW is almost always built with identical adhesives and dielectric films in these layers. Consequently, a given FPW dielectric system boils down to one film, one adhesive.

Adhesives summary

Adhesives are the weak but unavoidable link in FPW. Layer-type assembly technique requires an adhesive, even where PTF or covercoat technique are used: something must hold the structure together. [In polymer thick film (PTF), it's the binder in the ink and covercoat.] Adhesives largely determine FPW electrical, thermal, and chemical performance and are a critical factor in controlled impedance circuitry since adhesive is the dielectric which surrounds the conductor layer and, in typical constructions, amounts to 50% (or more) of the thickness.

FPW is classified by the dielectric film that's used—polyimide, polyester, and so forth—but the adhesive dominates performance and should be the main descriptor. Examples:

- Because they surround the conductors, adhesives are the primary insulation in FPW and determine electrical properties such as volume and surface resistance, dielectric constant, and dissipation factor. Film properties are secondary.
- Flammability is adhesive-driven. Virtually all films are self-extinguishing; most adhesives support combustion unless they're formulated to include suppressors.
- Adhesives are the controlling factor in critical properties such as elevated temperature insulation resistance and Z-axis expansion.
- Bond strength is an adhesive property: resistance to termination damage at assembly is determined by adhesive choice.
- Chemical resistance is also an adhesive function. It's quantified by change in peel strength after exposure to the test chemistry.

Films summary

Film properties determine the dimensional stability of the laminate which supports in-process FPW through etch and coverlayer (or covercoat). Laminate stability in turn controls production yield, thus the film has an effect on FPW cost which goes well beyond its purchase price.

It's fair to say that, except for specialized applications, all popular FPW films have adequate chemical resistance and electrical performance and are workably adherable. In almost every case, selection is based on cost and ability to survive FPW-to-hardware assembly technique, where harshest stress is applied. If soldering or other methods that involve temperatures above 150°C will be used, then polyimides or similar high-temperature films are the best choice; if lower-temperature methods—or very careful soldering with adequate heat shielding—are intended, polyesters, vinyls, or other low-temperature films can be used.

Adhesives

History

At the outset of its commercialization, FPW was built by fusion bonding. In this process, the base dielectric film is bonded to the foil, and in turn the coverlayer is bonded to the etched pattern, by application of enough heat and pressure to cause film melting. No separate adhesive is used—the process produces a homogeneous dielectric, of equal performance throughout. The fusion process allowed production of long cables by a technique called *hitch-feed lamination,* in which a series of overlapping lamination cycles was performed to create a circuit much larger than the press platens. Remelt of the overlap areas smoothened out edge-of-platen distortions because the dielectric was thermoplastic: in modern thermosetting adhesives, these areas would be forever damaged.

Popular dielectrics for fusion process included Kel-F, FEP Teflon,* polyethylene, polyester-polyethylene composites, and electrical-grade vinyls. For examples of fusion-bond circuits see Figs. 5-1 to 5-5.

Two-layer composites of polyester-polyethylene and FEP Teflon-

---

*Registered trademark of DuPont Company.

Figure 5-1 A 50-ft-long cable for a military application. The insulation is polyester-polyethylene composite. It was made by the print-and-etch technique with skip-screen printing.

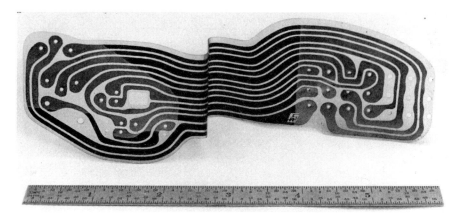

Figure 5-2 Early circuitry produced by roll-to-roll processing, used for telephone timing equipment. Manufactured in high volume on fusion vinyl/copper laminate with press-laminated vinyl covercoat. Conductors have a tin-lead finish. Note the application of the area-bare concept. (*Courtesy V. F. Dahlgren.*)

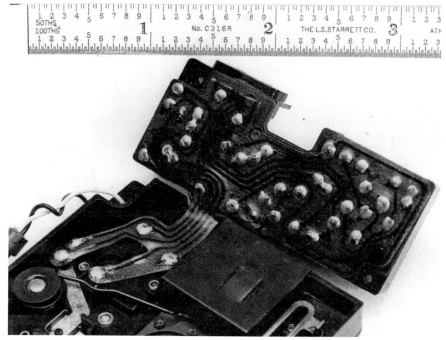

Figure 5-3 An early camera application. The shutter is connected to the automatic exposure module by vinyl-insulated flex circuitry. Note the solder attachments. (*Courtesy V. F. Dahlgren.*)

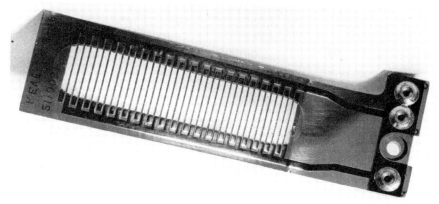

Figure 5-4 Gold-plated fuel-gage sensor. The dielectric is fusion FEP Teflon and polyimide. The terminal block is soldered in place. (*Courtesy V. F. Dahlgren.*)

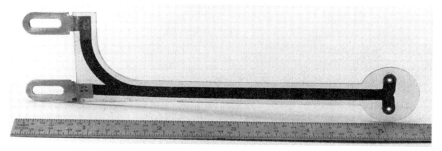

Figure 5-5 High-voltage two-layer fusion-bond Kel-F ignitor cable with spot-welded tabs at the left end and soldered eyelets at the round head end. (*Courtesy V. F. Dahlgren.*)

Kapton* were used because one of the layers melted at a lower temperature to function as a thermoplastic adhesive while the other, with higher melt temperature, afforded some degree of dimensional stability. Since the nonmelting film was the outside skin of the dielectric, this technique also eliminated short circuits to ground, a bizarre failure mode resulting from conductors pushing through the surface of molten dielectrics in single-polymer fusion coverlayering.

Thermoplastic polymers melt at a certain temperature and resolidify when cooled. The process is endlessly repeatable. Thermosetting compositions melt only during the first heat-up cycle, and subsequently won't

---

*Registered trademark of DuPont Company.

remelt, although some softening (above the glass transition temperature T_g) always occurs. The logical next step in FPW development from a two-layer fusion technique was a film-plus-adhesive process. This technique, because it is based on thermosetting adhesives, allowed lamination at even lower temperatures with the added advantage that the base layer adhesive didn't remelt when the coverlayer was applied.

Adhesives for early production use were developed specifically for the then-popular single-sided, pad-terminated FPW which required good peel strength, low flow, and easy processability. Low flow is necessary at pad apertures but is otherwise undesirable because it forces use of thicker adhesive layers to seal around etched conductor patterns. Increased circuit thickness, higher residual stress levels, and a dielectric system with a higher percentage of adhesive are the result of low-flow, bare-pad-adapted adhesives.

No FPW adhesive system, then or now, matches the mechanical, thermal, electrical, and chemical performance of high-performance films like the polyesters, polyimides, or Teflons.

New adhesive formulations and changes to established recipes occur often in FPW as material suppliers seek the best combination of properties. The following section is based on analysis of currently available materials. Considering the rapid changes which occur in the FPW industry, this discussion should be considered as a broad, generalized guide to adhesive properties and their effect on FPW performance.

FPW adhesive thicknesses range from 0.0007 to 0.003 in. The ability of low-flow adhesives to encapsulate conductors is limited; common practice is to deform the coverlayer film around the runs to reduce the amount of adhesive required to seal them. By this means, 0.001 in of adhesive can encapsulate 0.0014-in runs; nevertheless, even with the minimum practical 0.00075-in adhesive thickness in the base laminate, an FPW built with 0.001-in dielectric films contains 0.00175 in of adhesive, more than 50% of the 0.00375-in dielectric thickness.

Today, low-flow adhesives are no longer always required, because PTH termination—which eliminates coverlayer apertures, thus the need for low flow—is used in more than 50% of FPW. In rigid-flex circuitry (see Chap. 9), a complex, multilayered product with plated throughhole (PTH) termination, adhesive performance is of paramount importance; here the trend is to eliminate traditional FPW adhesives and replace them with more highly cross-linked adhesives or PWB-style prepregs which provide higher levels of flow, better thermal stability, and reduced residual stress. Increasing use of "adhesiveless" materials in rigid-flex as well as in FPW is a strong signal of the overdue recognition that traditional adhesives are the weak link in FPW dielectrics. (*Adhesiveless* in this context means absence of traditional, highly modified adhesives.)

Forms

FPW adhesives are used in several forms: as a single-sided coating on a film, as a double-sided coated dielectric film called a *bondply,* or as an unsupported or cast film:

- Base and coverlayers—identical in composition if not thickness—consist of single-sided adhesive coatings on a dielectric film and together form the dielectric system of single-sided FPW.

- Bondplies and cast-film adhesives are used to join layers of FPW together into multilayer structures. Bondplies provide an assured dielectric barrier between facing etched circuits, acting as a coverlayer for both; cast adhesive allows thinner constructions. Both are frequently apertured or outlined and registered into multilayer layups so as to confine the bonding function to desired areas—for instance, rigid regions of rigid-flex circuitry (see Chap. 9).

Coating processes. FPW adhesive products are produced from solutions of solvent, polymer, and curing agent. This mixture is machine-coated onto one or both sides of a film to form base, coverlayer, or bondply or onto a temporary carrier film to create cast-film adhesive. The solvent is removed in a drying oven, leaving a tack-free coating which is protected by a release film.

The release film is introduced at the last rewind stage and remains on the adhesive coating until layup. The release film is sometimes removed before drilling, in order to release any stress which could upset dimensions, then replaced after the operation. Stripping off this film creates a large static charge which contributes to foreign material buildup.

Requirements for adhesive coatings

- Adhesive coatings must be as tack-free as possible to minimize foreign material pickup and to allow easy registration and alignment of coverlayers and bondplies for lamination.

- Cast films—used in fragile 0.001- to 0.002-in thicknesses—must be sufficiently cohesive (but not brittle) to be lifted from the carrier and positioned in a layup without distortion. Most cast films are limp and weak, making accurate alignment of drilled or punched apertures with an etched pattern extremely difficult and labor-intensive.

- Coatings must have low residual volatile content (less than 1% is desired) to minimize lamination voids and poor adhesion. For the same reason, curing should not release volatile by-products.

- Cured adhesive must be resistant to attack by manufacturing process chemicals; as a conflicting requirement, drilling residues must also be removable from pad edges for the PTH process.
- Because they're all thermosetting to some degree, it's good practice to protect adhesive-coated inventories from elevated temperatures; refrigerator or freezer storage is preferred.
- Complete cure in a reasonable time at moderate temperatures aids production; 1 hour at 180°C is typical.
- All have limited shelf life and must be monitored to assure freshness and expected performance.

Adhesive types

Polyester. Used with polyester films, this primarily thermoplastic adhesive is roll-processable for high throughput at low cost. It can be made fire-resistant to meet Underwriters Laboratories specification UL 94VTM-0 when copper-clad. It processes at 75 to 100 psi at 140°C. Peel strength is around 5 pli. It will not pass the solder float test. Maximum service temperature is 110°C.

Butyral-phenolic. Used in high-flexural-endurance circuitry. A stiff, higher cross-link-density adhesive which provides uniform conductor support, minimizing deviation from the neutral axis. It is relatively fast curing—30 to 60 minutes at 176 to 148°C at somewhat high 250 to 400 psi pressures. Has modest but stable peel strength of 6 pli and resists exposure to 200°C for over 100 hours.

Modified epoxy. A class which includes a wide range of formulations and performances. Most cure at 177°C in 45 minutes to 1 hour at 1 to 300 psi. Peel is adequate at 10 to 12 pli. Has low moisture absorption and coefficient of thermal expansion (CTE), desirable in multilayer constructions.

Acrylic. The most commonly used adhesive for polyimide film systems. Available in FR versions that meet UL VTM-O with copper cladding. Provides excellent peel strength of 12 to 20 pli. Cures in 1 hour at 180°C. Moisture absorption tends to be greater than 2%. {Peel strength of all adhesives which absorb moisture may be affected by moisture content; degree of change from moist [70% relative humidity (RH) to dry (10% RH) conditions for one example was reported to be 6 pli, or 50%, with larger peel strength at high humidity.}*

Acrylics, as shown in Figs. 9-20 and 9-22, can have extremely large

---

*John A. Kreuz, "Kapton Polyimide Film in Flexible Printed Circuits," DuPont, 1977.

CTEs, particularly at elevated temperatures required for solder attachment and MIL-spec product qualification. In restrained constructions such as rigid-flex circuitry with PTH terminations, large expansion can result in dielectric separations and product rejection.

Summary of desired adhesive properties

An adhesive should have the following properties:

1. Coatable, with adequate shelf life
2. Low tack, low volatiles (uncured form)
3. Excellent adhesion to foils and films with controlled flow:
    a. Low for FPW
    b. Higher for PTH and multilayer constructions
4. Resistant to process chemistries (but desmearable)
5. Good electrical performance: low dielectric constant K and dissipation factor; high volume, surface, and insulation resistance
6. Low CTE; high glass transition temperature T_g, the practical upper temperature limit
7. Short, relatively low temperature curing
8. Low sensitivity to moisture absorption

Careful comparison of the properties of dielectric films (Fig. 5-7) with those of laminates (Fig. 5-8) and coverlayers (Fig. 5-9) illustrates the importance of adhesives.

Films

There are many requirements for films. To withstand production process stresses, FPW films must have high tensile strength and tensile modulus, high melting point with good thermal stability, low residual stress, low CTE, and high T_g. In finished FPW, desirable film properties include low dielectric constant and dissipation factor with high volume and surface resistivities.

Films should be available at acceptable cost (preferably from multiple sources) and should be transparent (to conform to industry expectations). They should have low moisture absorption (less than 1%) and an adherable surface to aid both production of FPW and assembly operations such as marking and potting. They should be machinable by drilling, routing, or die cutting.

Descriptions

The lowest-cost film in general use is polyethylene terephthalate (PET), also known as polyester. Roughly 20% of U.S. FPW production, some $70,000,000 yearly, is in polyester laminates and coverlayers.

Available from several major sources, this film provides excellent thermal and mechanical performance and aids in production of low-cost FPW both because the film is inexpensive and because it allows roll processing with thermoplastic adhesives. The primary factor limiting wider polyester use is its relatively low melting point of 250°C and low T_g of 80°C which make mass solder termination a difficult task. Polyesters stiffen at low temperatures.

Polyethylene naphthenate (PEN) is a close cousin to polyester which has not yet penetrated into FPW to a significant degree. Its performance is similar to PET but with improved thermal properties, including a T_g 40°C higher and a melt temperature 12°C higher.

PET costs around $3 per pound; PEN sells for $13; both are in contrast to polyimides, which command premium prices to $80 per pound. As mentioned, use of PET is limited by thermal performance which doesn't allow high-yield mass soldering. The added 12°C melt point of PEN may be enough improvement to stimulate significant usage, in which case production cost will drop by half to become roughly twice that of PET.

PEN is produced in a highly crystalline form through biaxial orientation. The area increase is about 10×; for example, 1-mil films are extruded at 10 mils and drawn to finish gage.*

Polyimide films are used in most—nearly 80%—U.S.-produced FPW. These relatively high-cost but high-performance films of choice provide an excellent blend of properties and are written into most FPW standards and specifications.

The largest U.S. polyimide film producer is DuPont Company, which markets several different films under the tradename Kapton®. Another major source is Ube Corporation, which sells polyimide film in the United States as Upilex®. *Polyimide,* like *epoxy,* is a generalized description of part of a chemical structure and should not be considered definitive of performance.

Commercially available polyimide films have moisture absorptions ranging from less than 1% to over 3%, with tensile properties that range from "elastic" to "stiff" and surfaces that are very adherable—or relatively not adherable. Some polyimides are functionally thermoplastic, even used as adhesives; others have unknown (because high) melting points. Consequently, it's very important to fully specify which polyimide film is used in FPW materials. Examples:

- The coefficient of hygroscopic expansion (CHE) for a once-popular polyimide film was reported to be 22 ppm/% RH. If the relative

---

*PEN laminate is available from GTS Flexible Materials Inc., Warwick, RI 02886.

humidity changed by 20%, the resulting change in dimensions for this film was 0.44 mils per inch, or 0.0044 in in 10 in, with equilibration to humidity occurring rapidly and reversibly.

- Another widely used polyimide expanded by 0.010 in per inch—1%—upon exposure to a common photoresist stripper.* By constant study of industry concerns and improvement at the molecular level, polyimide film suppliers have evolved versions of this workhorse product which are considerably more stable against process and use environments.

Other films

Limited use is made of other dielectric supports and films in specialty applications. Examples are aramid papers, fusion-bonded FEP Teflon, electrical-grade vinyls, and a range of flexibilized epoxy/saturant systems.

Aramid papers aren't films, in the strict sense, but function as a mechanical backbone and adhesive carrier just as do polymer films. Their advantages are extremely good dielectrical performance (dielectric constant $K = 1.6$ to 2, dissipation factor = 0.0015), good dimensional stability, low cost, and availability in a range of thicknesses. Disadvantages are high moisture absorption and a tendency to stain and discolor as process residues absorb into the fibrous structure. There are both United States and Japanese sources for aramid papers.

Fusion-bonded FEP Teflon (and polyethylene) circuitry provides best electrical performance but is difficult to produce in quantity at low cost. The reasons are the same as in early days of FPW production: the fusion process has poor dimensional stability, particularly during coverlayer lamination when the base layer is partially melted and flows under lamination pressure.

Electrical-grade vinyls are seldom seen today but have been used to produce high volumes of low-cost circuitry by the roll-to-roll technique. Usable to 105°C, solderable with care, vinyls are excellent low-cost insulators. Dielectric properties ($K = 4.7$, dissipation factor = 0.093) are satisfactory for interconnect use; dimensional stability is poor but adequate for simple, large-feature circuitry.

Where dynamics and flexure aren't needed, systems consisting of a saturating mat with flexibilized epoxy or other polymer should be considered. These materials can be thought of as PWB substrates with enough flexibility to allow limited bending and shaping for FPW-like use. Advantages are low cost with adequate performance and avail-

---

*John A. Kreuz, "Kapton Polyimide Film in Flexible Printed Circuits," DuPont, 1977.

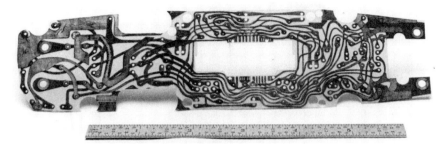

Figure 5-6 Early circuitry used in a telephone handset. The dielectric is nonwoven fiber mat with flexibilized epoxy. The wiring is double-sided with plated throughholes. Throughholes and outline were punched. The surface finish is nonreflowed tin-lead. (*Courtesy V. F. Dahlgren.*)

ability in roll form for high-volume production. See Fig. 5-6 for an example of high-volume telephone circuitry in epoxy-mat dielectric.

Materials based on aramid papers with FR epoxy or other thermosetting dielectric resins are available. Although the minimum substrate thickness (0.004 in) is greater than for polyimide films, these materials have been used to build prototype multilayer circuits with coextensive "flexible" layers (so-called rigid-flex; see Chaps. 7 and 9). Advantages of aramid papers are that their X-Y CTE is very low (although the Z CTE is correspondingly quite high), thus providing very good dimensional stability for layer-to-layer alignment, and they are easy to drill and PTH-process because there are no glass fibers. The dielectric structure approaches homogeneity (because the fibers are very small and evenly distributed) and, with FR resin, the system is nonflammable.

Summary of desired film properties

1. High mechanical stiffness—tensile strength and modulus
2. High operating temperature with low CTE
3. Good tear strength
4. Processability—machinability and adherability
5. Low moisture absorption
6. Low cost, readily available

Details of the properties of popular FPW dielectric films are set forth in Fig. 5-7.

Dielectric Materials

Property	Polyimides		Upilex S	Polyester	PEN
	Kapton				
	FPC-V	FPC-E			
Tensile stress, kpsi	34	40	55	28MD (machine direction) 33TD (transverse direction)	32
Modulus, kpsi	400	800	1238	600MD 700TD	870
Initial tear strength, g/mil	1090	1090	575	1000	1000
Propagating tear strength, g/mil	12	12	8.25	16MD 12TD	11.5MD 12.5TD
CTE, ppm/°C (20–100°C)	24	17	10	23MD 18MD	20TD 21 TD
CHE, ppm/% RH	17	9		10	10
Shrinkage, %	0.03 (200°C)	0.03 (200°C)	0.07 (250°C)	2–5 (195°C)	<0.8 (195° C)
Moisture absorption, %	3	2.4	0.9	<1	<1
O_2 permeability, $cm^3/m^2/day$ (3-mil film)	114	4	0.8	1400	500
Dielectric constant*	3.2	3.3	3.5	3.3	3.3
Dissipation factor	0.004	0.004	0.0013 (0.0078 at 200°C)	0.002–0.016	0.002–0.016
Volume resistivity, ohms	3.6×10^{17}	3.1×10^{17}	10^{17}	10^{18} (25°C)	
Surface resistivity, ohms	$10\text{–}13 \times 10^{17}$ (dry)	$>10^{16}$	10^{16}		
T_g, °C			>500	80	122
Melting point, °C	>400	>400	>400	254	266

*Dielectric properties vary depending on humidity, frequency, and temperature.

Figure 5-7 Representative properties of dielectric films.

Comparison Charts

Comparison between competing films, laminates, and coverlayers from manufacturers' data sheets is made difficult by the use of a wide range of test procedures and conditions. Nevertheless, information set forth in the following figures helps to convey the relationship among film, adhesive, and FPW performance.

The reader should understand that quoted data is as accurately reported as possible, but in some instances rounding and averaging have been used for clarity in translating manufacturers' data into these charts. It's also important to realize that manufacturers publish data in response to industry standards and specifications; other performance factors may be more significant to a particular application and should be discussed with material manufacturers for their assistance and further data. Lastly, new products are introduced and others withdrawn on short notice.

Comments on Fig. 5-7:

1. Upilex S has superior stiffness (tensile strength and modulus) and low CTE, leading to good dimensional stability. The high modulus results in lower propagating tear strength.
2. Low moisture absorption and permeability are desirable in multilayer constructions where trapped volatiles can cause lamination voids at high-temperature process steps.
3. Except for reduced operating temperatures, polyester, and PEN films are comparable to polyimides. PEN is a semiproduction film; it may become popular because it has better heat resistance than polyester and is much less expensive than polyimide.
4. Polyimide films are designed to have CTEs which match copper (17 ppm/°C). Adhesives have CTEs orders of magnitude higher (see Figs. 9-20 and 9-22).
5. Although important, chemical resistance is difficult to summarize in a table such as this, and seldom influences film choice. All films in current use have excellent resistance to common electronic manufacturing chemicals.

Comments on Fig. 5-8:

1. Variations in quoted data are slight.
2. Upilex FR system offers excellent performance.
3. There are no commonly employed tests of electrical performance at elevated (80°C) temperatures.
4. Resistivities are one to two orders of magnitude less than for films.

Property	Kapton-acrylic Standard	Kapton-acrylic FR	Upilex: Epoxy	Upilex: Butyryal-phenolic, standard	Upilex: Acrylic Standard	Upilex: Acrylic FR	Upilex: Epoxy mat
Peel strength, initial, pli	20	12	11	6	12	10	8
Peel strength, after solder, pli	10	11	10	6	12	10	
Dimensional stability, %*	0.1	0.1	0.11	0.07	0.035	0.01	0.08
Dielectric constant K	3.6	3.5	3.2	3.8	3.8	3.8	
Dissipation factor (1 MHz)	0.02	0.02	0.02	0.01	0.04	0.01	0.015
Volume resistivity, Ω (ambient temperature)	10^{15}	10^{15}	0.5×10^{15}	10^{14}	10^{15}	10^{14}	10^{13}
Surface resistivity, Ω (ambient temperature)	10^{14}	10^{13}	10^{11}	10^{10}	4×10^{11}	10^{10}	
Moisture Absorption, %	1–3		<4	1.7	5	1.7	0.4
Flammability		VTM-O				VTM-O	V-O (UL)
Solder float (10 seconds, 288°C)		Pass	Pass	Pass	Pass	Pass	Pass

*Dimensional stability values are composites from a variety of test procedures.

Figure 5-8 Properties of laminates from manufacturers' data.

5. Perhaps the most serious performance concerns with polyimide-film-based FPW dielectric systems are moisture absorption and high Z-axis expansion. These factors complicate production and qualification of multilayer FPW products. Newer film compositions and increasing use of more cross-linked, less flexibilized adhesives and adhesiveless laminates based on polyimide films promise simpler production with higher yield.

Comments on Fig. 5-9:

Film:	Kapton			Upilex S		
Adhesive type: Property	Acrylic	FR	Epoxy	Butyryal-phenolic	Acrylic	FR
Peel strength, initial, pli	10	9	11	6	12	10
Peel strength, after solder, pli	10	9	10	6	12	10
Dimensional stability, %	0.07	0.03	0.12	0.02	0.03	0.03
Dielectric constant K	3.6	3.5	3.2	3.8	3.8	3.8
Dissipation factor	0.02	0.02	0.02	0.01	0.04	0.01
Moisture absorption, %	—	—	<4	1.7	5	1.7
Flow, mils/mil	4.2	4	*	*	*	*
Volume resistivity, Ω	10^{15}	10^{15}	0.5×10^{15}	10^{14}	10^{15}	10^{14}
Surface resistivity, Ω (ambient temperature)	—	10^{13}	10^{11}	0.4×10^{11}	10^{10}	10^{10}

*Depends on press conditions.

Figure 5-9 Coverlayer properties from manufacturer's data.

1. Data was developed from laminates made by bonding foil to coverlayer, then etching to create test patterns.
2. There are no commonly employed tests of electrical performance at elevated (80°C) temperatures.
3. Resistivities of coverlayers are one to two orders of magnitude less than for films.
4. Flow into pad apertures is an important property for direct-terminated FPW but is strongly influenced by lamination practice (see "Coverlayer Process" in Chap. 8). There are established test procedures for measuring flow; a typical standard is 0.001 in of flow per 0.001 in of adhesive thickness.

Potting

The best possible protection for termination areas is potting, or encapsulation, of the junction of FPW and hardware in a cured-in-place block of insulation. This technique fully supports the terminals and protects them from mechanical or environmental stress. It is frequently used in military and high-reliability commercial assemblies.

It's desirable for a potting compound to be tough and somewhat resilient, rather than rigid, to avoid wear and cutting hazards at the junction of FPW and potting. Both thermosetting and thermoplastic polymers are used; thermosets include epoxies, polyurethanes, blends of epoxy and polyurethane, and silicone rubbers.

Thermoplastic materials offer very short cycle times and low cost for high-volume applications which are suitably tooled to allow injection-molding techniques, but elevated temperatures and pressures are required, even for low-melting compounds. Careful mold design, choice of compound, good sealing between mold and FPW, and protection of the FPW from the hot compound are all critical considerations. See Chap. 11, "Assembly," for further discussion of potting techniques.

Hardware

Connectors and other hardware to be potted must be sealed to prevent intrusion of the liquid compound into contact areas or onto mating surfaces. In connectors, one half of the pair typically has floating contacts while the other half has contacts which are locked and sealed

Figure 5-10 Rigid-flex assembly showing round MIL-style connectors, rectangular connectors, and PTH patterns used as connectors. Typical of military designs, this example has polyimide-glass cap boards supporting the PTH terminations. The high-temperature dielectric system easily resists damage at solder attachment and coating to insulate the solder joints. (*Courtesy Parlex Corp.*)

into the connector body. Here is a rare instance of things working out for the best, because potting onto the sealed half is best for potting purposes and correct for connector function: the compound won't leak through the rigid half and won't destroy the floating contact alignment. If the interface is already fixed by design and forces use of the floating half on FPW, the best technique is to hand-seal each contact with resilient compound, then pot while mated with the other connector half to hold the contacts in correct alignment.

Potting can be used to hold the FPW in a formed or bent condition for improved fit into tight assemblies.

Conformal coating

Conformal coatings are occasionally found in FPW assemblies as a topcoat over arrays of exposed solder joints, or in any areas which need protection but can't be potted. The purpose is to protect against accidental short-circuiting in these dense areas as well as to protect the body of the FPW from abrasion on the sharp fillets. Figure 5-10 is an example of conformal coat.

Summary

Conventional FPW dielectric systems consist of a support film and adhesive coating. The films, ranging from PET to polyimides and fluorocarbons, determine mechanical properties of tensile strength, dimensional stability, and stiffness. Adhesives have strong influence over electrical performance, because they're closest to the conductors. Both film and adhesive affect thermal behavior—CTE, maximum use temperature, termination technique. Conventional adhesives are formulated to have low flow and considerable flexibility, properties which are appropriate in bare-pad designs with premachined coverlayer apertures, but undesirable in multilayer constructions.

Potting compounds and conformal coatings are used to insulate and protect termination areas. Potting requires a mold and considerable process labor; conformal coating is faster and simpler.

Chapter 6

Conductive Materials

FPW is manufactured in a range of metal foils. The types of foil, ranked by volume of consumption, are

Wrought or rolled copper

Electrodeposited copper

Specialty alloys

Beryllium copper

Aluminum

Conductive inks or polymer thick film (PTF)

Copper Foils

Copper foil is far and away the leading choice for FPW conductors. Aluminum is superior in conductivity per dollar, but is chemically more active and therefore presents corrosion problems when joined to other metals at terminations. Overall, copper is preferred because it is the standard of conductivity and current-carrying capacity. It's easily etched to create high-density patterns, and can be deposited from solutions in multilayer or three-dimensional circuitry manufacturing processes. All standard methods for termination including a wide range of solders and braze alloys, thermocompression, and (with difficulty) spot welding can be used with copper.

There are two methods for converting copper into foil: electrodeposition (ED) and roll reduction. Industrial availability of copper and associated processing chemistry is unmatched; specifications, standards, and customers all contemplate copper conductors.

Electrodeposited foil

The ED process starts from an ionic solution and builds foil at high deposition rates on a rotating cathodic drum. It produces a foil which is smooth on one surface (the drum side) and rough on the other with vertical grain orientation and high yield and tensile strengths. High production rate and direct creation at the desired thickness lead to low cost; the vertical grain structure provides smooth, even etching properties to aid the production of narrow conductor runs on close centers. Because of their inherently rough surface, conventional ED foils bond with great strength to PWB dielectric systems, where their reduced flexural properties are irrelevant. Thin foils are least expensive if produced by the ED process; in ED, cost rises with increasing thickness.

Rolled foil

Rolled foils are different in every way. Because this process starts from ingots and has a longer production cycle with incremental approach to final thickness, rolling results in a foil with horizontal grain orientation, smooth surfaces on both sides, and a cost which rises with reducing thickness. Figure 6-1 is a cross-sectional photomicrograph which shows fine, horizontally oriented grain structure of rolled, or wrought, copper foil.

Figure 6-1 Detail of rolled-foil conductors in Zontar RF cross section. Grain structure is visible in the rolled foil and shows horizontal orientation. Etchback is improved because prepreg surrounds the pads.

Roll reduction quickly causes work hardening with the result that rolled foils must be periodically annealed. They are sold in several degrees of hardness, from *as rolled* to *dead soft*. Rolled-annealed (RA) is a standard anneal condition which provides good flexural endurance and resistance to fracturing in the dynamic use which is typical of FPW. FPW engineers automatically specify RA foil for FPW; most flexible laminates are manufactured with RA foil and most circuit specifications require it.

Modern electrodeposition technology has powerfully altered the logic of choosing RA foils for FPW. Through careful adjustment of current density and bath chemistry and by use of unique process equipment, high-quality ED foils are produced which have competitive if not superior flexural properties for FPW use. Figure 6-2 shows the fine, horizontally oriented grain structure of modern ED foil as found in chemically deposited metal-clad adhesiveless materials.

Because of the brittleness of early ED foil, it has long been customary to use RA foil in FPW materials. RA metallurgy provides good flexural endurance in FPW, but it also complicates handling during manufacture because it is soft and easily deformed. An alternative form of rolled foil offering better handling with good flexural properties is a version called *low-temperature annealed* (LTA) foil. LTA foil

Figure 6-2 Close-up of adhesiveless layer-barrel interface. Horizontal multilayer grain structure in high-endurance ED foil is visible. Near-vertical pad edges show excellent etching properties. Coverlayers are prepreg-bonded. Etchback is inhibited by the absence of an adhesive layer beneath the foils. The electroless-plate interface is heavily stained.

is sold with high levels of residual rolling stress (i.e., the grain structure is elongated to an unusual degree); high residual stress means quicker annealing when the foil is heated. LTA can be a good foil choice which resists handling damage because it initially has very high yield strength and therefore resists denting and creasing, but anneals during coverlayer lamination to a softer, more flexurally forgiving grain size. However, levels of residual stress which are needed can lead to instability and variable behavior including unwanted preannealing, resulting from storage at excessive temperatures and leading to unexpected handling damage, or excessive annealing at lamination.

A table of copper foil properties extracted from IPC-CF-150E, published by the Institute for Interconnecting and Packaging Electronic Circuits, follows:

Copper type and class	Thickness, mils	Tensile strength, kpsi	Ductility CHS, %	Fatigue ductility
Electrodeposited foils				
1	0.7	15	2	—
	1.4	30	3	—
	2.8+	30	3	—
2	0.7	15	5	—
	1.4	30	10	—
	2.8+	30	15	—
3	0.7	15	2	—
	1.4	30	3	—
	2.8+	30	3	—
4	—	—	—	—
	1.4	20	10	—
	2.8+	20	15	—
Wrought or rolled foils				
5	0.7	50	0.5	30
	1.4	50	0.5	
	2.8+	50	1	
6	1–2+	25–50, depending on temper	1–20	30–65
7	0.7	15	5	65
	1.4	20	10	
	2.8+	25	20	
8	0.7	15	5	25
	1.4	20	10	

Note: By agreement, 0.7 mil or 0.0007 in is $\frac{1}{2}$ oz; 1.4 mils or 0.0014 in is 1 oz; 2.8 mils or 0.0028 in is 2 oz. Ounce classifications—the weight of a square foot of foil—are obsolete and derive from sheet-metal and roofing use.

Class 7 is RA, generally used in FPW. Class 8 is LTA foil.

Foil	IPC class	Ultimate tensile strength, kpsi	Elongation, %	Cycles to failure, 0.078-in-diameter mandrel
Rolled foils				
As-rolled	5	60	2	200
RA	7	25	30	275
LTA	8	18	35	300
Electrodeposited foils				
ED, JTCS	3	60	17	115
Annealed JTCS		32	30	300
Annealed ED JTC AM		52	34	469
NelFlex[†]		43	24	298

*Reproduced with permission from the *Proceedings of the First International Conference on Flex Circuits* (FLEXCON TM 94). Copyright Semiconductor Technology Center Inc., P.O. Box 38, Neffs, PA 18065.
[†]NelFlex is a trademark of Nelco International Corporation.

Figure 6-3 Flexural endurance test results.*

Practical flexural endurance testing of FPW produces interesting results, as indicated in Fig. 6-3.

Figure 6-3 shows that LTA, annealed ED foils (JTCS), and NelFlex materials all have excellent flexural endurance, in this test better than RA (respectively 300, 300, and 298 versus 275) in cycles to failure on a 0.078-in-diameter mandrel. It also shows that elongation is not a good indicator of flexural endurance. NelFlex is an adhesiveless material consisting of a sputter-deposited metal seed layer on polyimide film with electrolytic plate-up to the desired conductor thickness. Figure 6-4 gives the thickness equivalents of foil weights.

Nominal thickness, in	Tolerance, in	Weight, oz/ft^2
0.0007	0.0001	$\frac{1}{2}$
0.0014	0.00015	1
0.0028	0.0003	2
0.0042	0.0004	3
0.0056	0.0006	4

Figure 6-4 Typical foil tolerances (from IPC D-249).

Surface treatment

Copper is a highly reactive metal which slowly oxidizes under room storage conditions and more rapidly misbehaves at lamination temperatures. "Raw" (untreated) copper therefore doesn't adhere well. To improve the stability and adherability, particularly of rolled foils, a wide range of surface treatments are used, starting with formation of oxide films and proceeding through intentionally rough ED copper to brass and bronze platings.

The bulk of RA foil used in FPW is treated on one surface. A special form, called *double-treat,* which is passivated on both surfaces, is offered as a performance enhancer for image and coverlayer lamination. Unlike single-treated foils which must be pumiced (see Figure 6-5, which shows a hand-pumice station) or chemically cleaned before resist cladding, this foil doesn't require precleaning—it goes directly into the clean room for resist cladding. After etching and resist stripping, the surface is simply rinsed off, dried, and transported to makeup for lamination. Solder wicking under the coverlayer (see Chap. 8) and lamination voids in the A zone after thermal stress are sharply reduced by double-treated foils. Cost of double treating is not trivial, but the offsetting benefits of eliminating two cleaning cycles plus enhanced adhesion and stability are very significant.

Figure 6-5 Hand-pumice station has air-driven rotary brush, rinse hose, and black elastomeric support that cushions and grips the FPW sheet while it is cleaned.

Other Conductive Materials

Other choices, used in low volumes in specialty or developing markets, are specialty alloys, aluminum, and polymer thick film inks.

Nickel, nichrome, constantain, and other specialty alloys are used in FPW that interconnects with liquid–nitrogen-cooled circuitry. The reason for using these low-conductivity metals is to reduce conduction of heat into the nitrogen-cooled vessels; the accompanying lower electrical conductivity is compensated by increased amplification.

Aluminum

Aluminum is an excellent conductor and provides a weight savings, compared to equally conductive copper, of roughly one-third. Aluminum has less flexural endurance and cannot be terminated as easily, although specially prepared terminal areas may allow solder attachment. For chip-on-FPW use, semiconductor termination techniques (such as wire bonding) are more easily applied to aluminum than to copper.

Beryllium copper

Beryllium copper is used for special FPW where extended flex life is critical, and where formed or etched-in spring configurations provide self-terminating conductors. Figure 6-6 shows a jumper-like FPW with beryllium-copper conductors in flat and formed condition together with

Figure 6-6 Forming fixture with unformed and formed U circuits in beryllium copper.

Metal	Resistance, 10^{-6} Ω/cm	CTE, ppm/°C	Specific gravity	Tensile strength, kpsi	Elongation, %
Copper, RA	1.724	16.6	8.89	35	1-20
Copper, ED	1.776	16.6	8.94	50	1-12
Aluminum	2.83	23.6	2.7	12	18
Nickel	9.5	13.3	8.9	72	40
Stainless steel	91	17.3	7.9	90	50

Figure 6-7 Metal foil properties.

the simple forming tool. This circuit clips over the edge of a double-sided PWB to provide direct interconnection between topside and bottomside terminals. It eliminates the plated throughhole (PTH) process.

Foil properties

Figure 6-7 lists electrical, thermal, and mechanical properties of several kinds of metal foil.

PTF and Conductive Inks

PTF inks enjoy rising popularity based on low cost, reduced environmental concerns, and simplified manufacturing. PTF has significantly lower conductivity (about one-tenth that of bulk copper or 30 mΩ/sq). Because they are much thinner (0.0002 to 0.00025 in total), screen-printed PTF conductors have about one-thirtieth the conductivity of the same width run in 1-oz (0.0014-in) copper. For applications where high conductivity isn't relevant—semiconductor interconnects, keyboards, displays, and the like—PTF is a good choice. Pattern density is limited to screen-print capability, about 0.008-in lines and spaces.

Where significant current-carrying capacity in some runs is required, PTF circuitry can be combined with etched copper in a composite circuit with etched runs on one surface of a dielectric and PTF runs on the other.

Interconnection between PTF runs and other PTF layers or components or hardware to form assemblies is normally achieved by conductive adhesive, an emerging technology which has great promise for process simplification and cost reduction. Provided long-term reliability issues are resolved, conductive adhesives and direct interconnection by "wet overprint" methodology promise mass termination at low cost.

Multilayer constructions in PTF technique are easily created with dielectric overprint on a conductive pattern, with suitable apertures where interconnection is desired, to insulate between conductor layers, which are added, one atop the other, to the desired level.

A major advantage of the PTF process is that it sharply reduces thermal demands on the FPW substrate, because elevated process temperatures aren't required either to build the circuit or to terminate it. This simplification allows use of very low cost, high-performance but thermally limited dielectric films such as polyesters, PEN, and Ultem.*

Acceptance of the unique fabrication and termination methods, resolution of issues of conductivity and reliability, and density limitations imposed by the screen print process are the only barriers to rapid penetration of this technique into the interconnect market.

Shielding Materials

Shield is a verb and a noun, meaning both the actions taken and the materials employed to protect circuits from disturbing electrostatic fields. The simplest form of FPW shielding is grounded adjacent conductor runs. Application of conductive layers with associated connection to a ground circuit is more commonly used; analogous in wire and cable is a braided or foil-wrap outer jacket which isolates a cluster of wires as a group. In FPW, more complex shielding designs are possible: the shield layer may be etched or otherwise broken up into multiple, independent shields so that individual runs are separately shielded where common ground currents are undesirable.

A variety of materials has been used for shielding:

- Copper foil, with or without etched apertures
- Conductive (PTH) screened layers, with or without apertures
- Electroless deposition
- Woven screen or mesh
- Self-adhesive metal foil tapes

Material choice depends on the criticality of several factors:

- Shield effectiveness
- Impact on flexibility
- Cost
- Durability
- Method of termination

*Registered trademark of General Electric Company.

Solid copper foil provides the highest level of isolation but affects flexibility—it takes more force to bend a shielded FPW—and reduces flexural endurance or durability because the neutral axis is offset from either conductor or shield (both are compressed and elongated as the circuit flexes).

Other materials provide intermediate performance. Screen and etched-aperture constructions are used because they don't stiffen the FPW as much and perhaps have greater durability compared to solid foil. Apertured designs are less effective to the degree that disturbing field energy can enter through the apertures. See Chap. 4 for more information on shield effectiveness.

PTF shield layers are screen-printed directly onto the FPW surface or may be printed on a coverlayer which is later laminated onto the FPW. An advantage of the two-step technique is that the PTF pattern may be applied to the adhesive side of the coverlayer, which then acts as an abrasion protector for the PTF.

Summary

Copper foil is the most popular material for FPW conductors. It is produced by either roll reduction or electrolytic deposition; rolled-annealed foil is the usual choice.

Special surface treatment is applied to copper foils to passivate the surface and enhance long-term adhesion.

Other metals which find limited use are aluminum and beryllium copper. Aluminum has theoretical advantages in weight/conductivity, but is difficult to terminate without introduction of electromotive corrosion potentials. Beryllium copper is used where spring properties are required; an example is combined connector/FPW uses such as top-to-bottom edge interconnectors for printed wiring boards.

PTF inks are cost-effective for low-current, low-conductivity applications. Inexpensively applied by silk-screen printing onto low-cost films such as PET and terminated by pressure connectors or conductive adhesives, PTF enables production of low-temperature interconnection systems for benign-environment, cost-sensitive packages.

FPW can be effectively shielded against electrostatic fields by application of conductive layers. These may consist of copper foils, PTF screened patterns, conductive tapes, or screen mesh. Effectiveness is surprisingly high; it exceeds that of conventional woven wire performance at higher frequencies.

Chapter 7

Adhesiveless Materials

Introduction

Adhesiveless is an unfortunate term for a class of laminates—functionally correct, but misleading and technically inaccurate.

As discussed in Chap. 5, "Dielectric Materials," the weakest link in conventional flexible printed wiring (FPW) laminates is the adhesive layer, where lowest thermal, electrical, and chemical performance results from satisfying demands for easy processing, low cost, and flexibility. Starting in the early '70s, accelerating through the '80s, and reaching significant market volumes today, adhesiveless materials are more correctly described as *high-performance adhesive* materials, since some of them contain adhesive at the laminate stage and all require adhesives for coverlayer or multilayer lamination.

The impetus for adhesiveless development may have been tape-automated bond (TAB) circuitry or demand for better flex materials in rigid-flex circuit construction. Both these applications suffer reduced yield from conventional adhesives and both can well afford the added cost of adhesiveless materials, which are increasingly cost-effective with advancing technology and rising performance demands. Whatever the historic stimulus, these second-generation FPW laminates are superior to conventional adhesive-based materials in important performance areas: dimensional stability, coefficient of thermal expansion (CTE), moisture absorption, temperature resistance, plated throughhole (PTH) processability, flexural endurance, thinness.

The thermomechanical goal for these newest FPW materials is to duplicate the performance of rigid laminates. Unfortunately, compared to conventional materials, they're more expensive, at least at this stage, and therefore are used only where their rigid-material-like performance boosts product yield enough to offset their premium cost or where their ultimate thinness provides a technical edge.

Definition

A proposed definition is as follows: "Adhesiveless laminates are composites of conductive and dielectric layers with glass transition temperature T_g above 180°C and CTE less than 150 ppm/°C. They may be single- or double-clad; they are a binary structure of conductive and dielectric materials homogeneous in nature, very flexible with excellent electrical and chemical specifications. Adhesives, if used, strive to be identical in performance and composition with the dielectric."

It should be noted that adhesiveless construction is theoretically attractive in any dielectric. To date, most FPW material has been polyimide-based, but polyesters, polyethylene napthenate (PEN), and fluorocarbons have valuable attributes for some uses and may become available.

Application Benefits

For single-sided FPW use, adhesiveless materials bring better dimensional stability, thin metallization for fine feature generation, high flex endurance, and better electrical properties. In double-sided and multilayer FPW, added benefits are reduced CTE, a simplified PTH process, and minimum thickness buildup as an important aid to manufacture of higher layer count designs.

Construction Methods

Adhesiveless laminates are manufactured by any of three methods:

1. Coating a polymer onto a metal foil
2. Metallizing an existing film
3. Bonding foil to film by means of a polyimide-like adhesive

This is a technology that's developing from traditional FPW roots but to date doesn't have a formalized structure. At this point it's impossible to say which method is best or will dominate. Coating on foil produces a single-sided material based on a pretested, certified foil that can be in any alloy. Deposited film metallization provides an inherently double-sided construction of plated metal on a pretested, certified film. High performance bonding allows both foil and film to be certified (providing the advantage of easy specification acceptance).

Both coating on foil and deposited metallization are roll-to-roll techniques; the high-performance adhesive method, because of process conditions, is limited to sheet form but could be available through a roll process, given sufficient demand.

The first adhesiveless materials consisted of an amide-imide coating on copper foil. This system was relatively low-cost, thin, and available in a variety of exotic foil alloys, but its biggest attraction was an easily etched dielectric, which allowed creation of very precise apertures for circuit access. Because the polymer contracted strongly during cure—which required high temperatures—this intrinsically single-sided material had high residual stress and significant postetch shrinkage and was difficult to process because it curled very strongly unless racked and restrained.

Coating on foil

Amide-imide–coated laminates were developed for TAB circuitry, where thermal stability, easy dielectric windowing, and availability in roll form are required. Absence of an adhesive layer meant these circuits could be thermocompression-bonded and thermal-cycled without loss of terminal integrity or dielectric degradation. Eventually, handling difficulties, poor dimensional stability, and lack of a double-clad version forced amide-imide out of the market.

Development of polyimide polymers with CTEs matching that of copper and low-contraction cure has revitalized the coating-on-foil industry; the latest materials have much lower residual stress and low after-etch shrinkage. The straightforward manufacturing process leads to low product cost.

Two layers of coating-on-foil material can be bonded together in a back-to-back relationship to form a double clad. This approach has been practiced by offshore producers as a way to enter the double-clad market. The manufacturing process is inherently multipass, which tends to elevate product cost. Presence of two different polymers in the cross section—a violation of the proposed adhesiveless definition—complicates plasma treatment and production of uniform etchback.

Deposited metallization

Advantages of deposited metallization over both coating and high-performance bond methods are

1. Thinnest metallizations are possible.
2. With preperforated films, PTH is provided at no added cost.
3. Dimensional stability, because neither elevated temperatures nor pressures are required, is highest.

Thin metallizations—on the order of 2 to 3 μm, or 100 μin—are very popular for extremely fine line circuitry used in a variety of semiconductor packages such as MCM-L (multi-chip-module interconnects based on laminates), BGA (ball grid array) circuitry, and the like. In these applications, current-carrying capacity is unimportant but ability to generate conductors and spaces of 0.001 in or less is required. PTH process is needed with double-sided circuitry to sort out the dense routing and complex circuitry that come with this microtechnology, and deposited metallization provides this function inexpensively: vias down to 0.002 in in diameter can be produced by high-speed punching or holographic laser perforation, then metallized simultaneously with the surfaces. The unavoidable induction of residual stress in the dielectric layer (which results from the high temperatures and/or pressures required for curing a polyimide coating or adhesive bonding a foil to a film) is absent from deposited metallization process, which occurs under stressless conditions at near room temperatures.

Because in this process the foil is never handled as an unsupported sheet, deposited metallization can produce smooth, uniform, undistorted foil surfaces at any FPW-processable thickness, something which is nearly impossible by either coating or bonding methods because of handling problems.

Deposition of metal is practiced in two forms, differing only in methods for applying the seed layer: (1) sputtering (vacuum metallization or so-called physical vapor deposition) and (2) "wet" (electroless) deposition, both followed by electrolytic buildup to the desired thickness. A barrier or seed layer of oxidation-resistant metal is used in both methods to assure stable attachment of dielectric to the metallization. Classic seed materials are nickel, chromium, and mixed metal oxides. Some seed layers—nickel is a good example—force use of a special post-conductor-etch process, because they're resistant to conventional copper etching baths. However, the performance benefits which they bring—long-term resistance to thermally induced bond degradation and stable peel strength—are well worth the slight process complexity required for their removal.

High-performance bond

The high-performance bond process, except for the extreme temperatures (over 300°C) required, is a straightforward extension of normal lamination. Dimensional stability is a concern whenever high lamination temperatures are used, but careful choice of film and adhesive formulation and tight process controls keep after-etch shrinkage in the range of conventional materials.

FPW can't be built without adhesives in coverlayers, bondplies, and cast films, and manufacturers prefer material systems which offer the same adhesive throughout. The high-performance bond technique is the only current "adhesiveless" production technique which could lead to a homogeneous dielectric structure. It's unlikely that significant volumes of FPW will be built using an adhesive system which requires bonding at more than 300°C, as is the case with the current high-performance system; 200°C is a reasonable upper limit. The chemists who developed the high-performance bond adhesive may be able to produce similar performance in a lower-temperature adhesive; when they do, this will be a very attractive system for FPW use.

Rigid Flex

Adhesiveless materials earn their keep in products that require maximum thermal performance and dimensional stability with least layer thickness; examples are MCM-L, TAB circuitry, and rigid flex (RF). In TAB and FPW use, covercoating or coverlayering adhesiveless-based circuitry with conventional adhesive-bonded films doesn't significantly compromise performance because X-Y stability comes from the laminate and Z-axis expansion is not relevant.

Multilayer products require higher performance. They should be built with high-T_g, low-CTE laminates and laminating adhesives such as epoxy-glass and polyimide-glass prepregs. Adhesiveless materials are attractive alternatives to both thin-core PWB laminates and conventional FPW laminates for these applications. They're thinner than PWB materials both in foil and dielectric and thus allow finer lines and spaces and more layers in the same thickness. Although they have the thermal performance of rigid materials, they're as flexible and resistant to flexural fatigue as FPW laminates; they therefore bring the best properties of both FPW and PWB materials to rigid-flex use.

Rigid-flex is built in significant quantities. Market size is hard to determine but certainly exceeds $65,000,000 in United States production yearly. An important trend in rigid-flex production is to the use of PWB-style prepregs to assemble multilayer areas. This is a happy case where what is necessary is available and usable: thermoset resin/glass constructions have the T_g and CTE which are needed in multilayer PTH products, approach the thermomechanical performance of polyimide films (thus meeting the spirit of the adhesiveless definition), and are usable in stiff, nonflexing multilayers.

NOTE: RF manufacturing technology is a rich field of patent activity. The reader is cautioned to review issued patents before committing to a manufacturing technology. See Chap. 9, for details of rigid-flex manufacturing technology.

Benefits

Reduced thickness is a major benefit of adhesiveless laminates (made by coating and metallizing methods) over conventional materials. A moment's reflection will establish that adhesive layers account for a high percentage—roughly half—of the cross-sectional thickness of a multilayer FPW product; removing adhesive from the cross section has a powerful thinning effect that improves packaging efficiency. Even with minimum adhesive thicknesses, a single-layer 0.0014-in foil circuit with 0.001-in dielectric base and coverlayer films is 0.00375 to more than 0.005 in thick (depending on adhesive flow at coverlayer) and contains 0.00175 in or 35% adhesive. If the base laminate is adhesiveless, overall thickness drops by 0.00075 in (a change of 15%), and the percentage of adhesive drops to 23.5%.

Improved thermal performance is immediately appreciated in rigid-flex circuitry (see Chap. 9) and in other applications such as automotive underhoods where long-term elevated temperatures degrade and decompose conventional adhesives and foil-to-adhesive interfaces. Coating-on-foil and deposited metallization systems have the thermal longevity ("heat-proofness") of their polyimide substrates, and high-performance adhesive systems based on thermoplastic polyimides aren't far behind. The limiting factor in the use of all these materials is oxidation of the copper conductors and/or termination failure.

Comparisons

Adhesiveless materials have better performance, compared to conventional adhesive-bonded laminates, in the areas of thermal expansion, moisture absorption, dimensional stability, and electrical properties. These are fundamental benefits of intelligent materials engineering and are readily realizable in production use. They may not stand out in a comparison of data sheets, however, because materials suppliers tend to be conservative in their claims. Figure 7-1 compares specification ("slashsheet") minimum requirements for a popular conventional FPW laminate (acrylic) and the three types of adhesiveless material.

Comments on Fig. 7-1: these requirements are established by the IPC and constitute the minimums required for certification to the 241 document.

There are omissions: physical-vapor-deposited material is available in a range of cladding thicknesses well above 1 μm; polyimide used in cast and chemical-deposited materials is not specified but does influence performance. A wide range of mechanical, thermal, and electrical behavior is available in polyimide materials.

Manufacturer's data sheets provide more specific data, as Fig. 7-2 shows.

Figure 7-3 presents data on one brand of cast-on-foil material, both single-clad and double-clad.

Comments on Fig. 7-3: chemical resistance is claimed to be excellent, with 100% retention of properties after exposure to a wide range

Property	Acrylic (baseline)	Cast	Deposited Physical vapor	Deposited Chemical
Slashsheet no.	1	11	12	18
Peel strength, pli, (0.0014-in foil)	4	6*	5†	6(≥0.0014 in)
Propagation tear strength, (0.001 g/mil)	4	5	4	4
Dimensional stability, % (method B)	0.15	0.2	0.15	0.1
Chemical resistance, %	80	80	80	90
Dielectric constant	4	4	4	4
Dissipation factor	0.04	0.01	0.012	0.012
Dielectric strength, kV/0.001 in	2	3.5	2.5	2.0
Moisture IR (insulation resistance), Ω	10^8	10^9	10^8	10^{10}
Moisture absorption, %	6	3	4	4‡
Flammability	DBD	VTM-O	VTM-O	VTM-O

*>0.0014 in = thinnest foil specified.
†<1 μm = thickest foil specified.
‡A property of polyimide film, but film type is not specified.

Figure 7-1 Minimum performance requirements from Institute for Interconnection and Packaging of Electronic Circuitry (IPC) document FC 241 for Class 3 materials.

Property	Acrylic (baseline)	Cast A (Mitsui)	High-performance bond A (DuPont AP)	High-performance bond B (Shin-Etsu RAS33D42)
Peel strength, pli* (0.0014-in foil)	9–12	>6	13	8
Dimensional stability, %	0.05	0.04	0.04	0.06
Solder resistance (30 sec/315°C)	Pass	Pass	Pass	Pass
Dielectric constant	3.6–4	3.2	3.2	3.4
Dissipation factor	0.02–0.03	0.007	0.002	0.03
Dielectric strength, (kV/0.001 in)	4–5	5	6	3.3
Moisture absorption, %	1–3	1.2	0.83	1.5
Propagation tear strength, g/mil	6–11		11	20

*Peel strength is universally used to define the quality of bond between foil and substrate.

Figure 7-2 Comparison of product data sheets.

Property	Single-clad	Double-clad
Peel strength, pli	7.8	7.8
Tensile strength, kpsi	37	22
Propagation tear strength, g/mil	5–6 min.	10
CTE, ppm/°C (100–240°C)	20	20
Dimensional stability, %	0.01	0.03
Solder float	400°C/1 minute	380°C/1 minute
Dielectric constant	3.5	3.5
Dissipation factor	0.007	0.007
Volume resistivity, mΩ·cm	1.3×10^{10}	10^4–10^5
Surface resistivity, mΩ	5×10^{10}	—
Dielectric strength, kV/mil	7.5	2–3
IR (insulation resistance), mΩ	10^{10}	—

*Data courtesy Nippon Steel Chemical Company, Ltd., for Espanex, brand name for their adhesiveless laminates.

Figure 7-3 Cast-on-foil data.*

of industrial chemicals. The polyimide is attacked (85% retention) by exposure to 10% NaOH for 20 minutes at 25°C, suggesting easy windowing by chemical removal of the dielectric.

Figure 7-4 presents data on deposited-metal material.

Adhesiveless covercoats

The thermal resistance of adhesiveless materials would be wasted in a circuit that included a conventional adhesive-bonded coverlay, but when combined with polyimide covercoat systems (photoimagable or screen-printed), the resulting FPW is entirely "adhesiveless" and capable of superior performance in severe environmental conditions. Polyimide covercoat systems normally require an elevated temperature cure, which complicates the manufacturing process and may compromise dimensional stability. An example is SPI-200* a screen-printable polyimide covercoat. According to the manufacturer, cure requires roughly 10 minutes exposure to 130°C followed by 2 minutes at 160°C; 2 minutes at 200°C and a final 2 minutes at 270°C. The cured coating is claimed to resist solder float at 320°C and to retain excellent long-term insulation resistance, which remains above 10^8 mΩ after prolonged exposure to 200°C.

---

*A product of Nippon Steel Chemical Company, Ltd.

Property	Value
Peel strength, pli	6–9
Tensile strength, kpsi	36 MD (machine direction)
	27.5 TD (transverse direction)
Initial tear strength, g/mil	650
Propagation tear strength, g/mil	10 min.
Fatigue ductility, %	266 (out)
	315 (in)
Flex endurance, cycles	77,652 (out)
	50,756 (in)
Dimensional stability, %	0.06 MD
	0.07 TD
Flammability	94 VTM-O
Solder float	Pass
Film-dependent properties	
Chemical resistance	85–95
Dielectric constant	3.4
Dissipation factor	0.008
Volume resistivity, mΩ·cm	1.3×10^{10}
Surface resistivity, mΩ	1.3×10^{10}
Dielectric strength, kV/mil	5.48
IR (insulation resistance), mΩ	2.5×10^6
Moisture IR (insulation resistance), mΩ	10^4 (wet)
	3×10^7 (dry)
Moisture absorption, %	1–3

*Data courtesy Polyonics Corp., 112 Parker St., Newburyport, Mass.

Figure 7-4 Deposited metallization data.*

The ultimate covercoat for an adhesiveless FPW is a photoimagable polyimide, because the resulting FPW is truly "adhesiveless"—fully insulated and completely free of low-T_g, high-CTE, low–cross-link polymers. A developmental system which will be commercialized in the near future sounds like the perfect choice. Its characteristics:

- Fully exposes with 500 mJ/cm^2 in 0.001-in coatings
- Develops aqueously
- Withstands severe creasing
- Doesn't shrink on curing
- Cures at 200°C in 1 hour
- Low cost

Peel strength

It's an easy property to measure and, because it's numeric, appears to provide the basis for objective comparison. But meaningful test of peel strength requires much tighter controls than are used.

Peel strength test consists of applying a force to separate a strip of material of measured width into two components through a prescribed angle of separation; the results are stated as *force per unit width* (an unusual, illogical unit). Force required is influenced by:

- Radius of bend over which the materials separate
- Force required to bend the layers to that radius
- Thickness of layers
- Stiffness of the materials of the layers
- Location of the cleavage line
- Angle of peel

Varying peel test results are everyday events which bear witness to the difficulty of controlling and replicating peel test results.

Peel strength is not a good test for FPW laminates because peel is not a natural FPW failure mode. What's worse, overemphasis on peel test data leads to choice of the wrong product, because high peel strength and poor thermal performance are consequences of the same adhesive property—plasticization. Strong, high-modulus bonds typical of adhesiveless materials fail by stress concentration when peeled, while low modulus, highly plasticized adhesives yield, distribute the peel force, and appear to be stronger. Thermally stable bonds, like those shown by cast A materials in Fig. 7-3 and deposited metals in Fig. 7-4, are stiffer—desirable for thermal performance—but distribute the peel force over a smaller area and consequently peel at lower values.

A major materials manufacturer, Sheldahl Inc., performed peel and Z-axis tensile tests on acrylic and directly metallized FPW laminates as received and after 10 days at 85°C, 85% relative humidity (RH) and reported these results:

Laminate type	90° peel strength, pli	Tensile strength, kpsi	Tensile strength after 85°C/85% RH, kpsi
Acrylic	14	3	3
Deposited metal	8	10	6

The aging data shows that deposited metallizations have higher tensile strength even after aging.

Figure 7-5 shows the mechanics of peel and tensile testing.
Peel values are given in *force per unit width*. Implied in this unit is standardized length, because strength is always measured as force per unit area. Views A and B illustrate how peel test is affected by uncontrolled area: plasticized adhesives, as shown in view A, elongate and distribute the load over a large area and thus appear stronger. In adhesively stiffer systems, as shown in B, the length of bond which is tested is much shorter because the attachment elongates less before failure; less area sharing the load means lower measured force which is reported as lower strength.

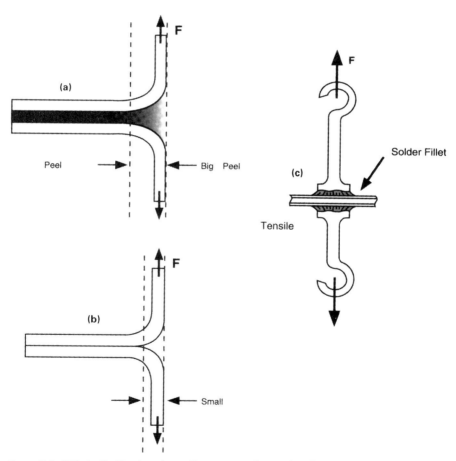

Figure 7-5 Effect of adhesive elongation on a peel test. An elastic, low-modulus adhesive (view A) extends the area under test and therefore appears stronger, compared to a stronger but lower-elongation adhesive (view B). A tensile test (view C) more accurately evaluates adhesive strength and better represents FPW loading.

A tensile test—application of force on an area of the laminate along an axis perpendicular to the plane of the material—is schematically shown in view C. A set of probes is soldered or bonded to pads etched in the laminate, then pulled apart. The area under test is the area of the etched pad; strength is force/area. Besides being physically logical and reproducible, this test mimics FPW failure modes.

CTE

Missing from data sheets is information important to FPW design and production such as CTE over the range from room temperature to 287.7°C, data on practical dimensional stability under factory conditions, and data on stability of bond with prolonged thermal exposure. FPW materials producers and industry groups attempt to characterize and control these products and publish data which is believed to be true and representative. In selecting materials on the basis of manufacturer's information, the reader should consider the difference between *certified* and *typical* product data. It is simple-minded to expect that product information is always correct, however; manufacturers should constantly test incoming laminates for these properties and choose vendors according to the results. See "Adhesive CTE" section and Figs. 9-20, 9-21, and 9-22 in Chap. 9 for TMAs of FPW adhesives.

Summary

Excessive thermal expansion in conventional FPW adhesives is avoided in an emerging class of insulated foil materials called *adhesiveless*. These materials are manufactured by a variety of techniques which eliminate low-T_g, high-CTE materials.

Other advantages of adhesiveless construction are improved dimensional stability, increased chemical resistance, lower moisture absorption, and better electrical performance.

RF constructions are enhanced by use of adhesiveless materials which improve yield in MIL-P-50884 thermal stress and thermal shock testing. TMA of RF structures built with these materials shows 50% reduction in Z-axis expansion at MIL-spec test temperatures.

The peel test, a widely used FPW quality standard, leads to erroneous conclusions when applied to stiff, high-strength materials which perform poorly in this ambiguous and irrelevant procedure.

Chapter 8

Manufacturing Processes

Introduction

All manufacturing organizations can be divided into two groups according to their dedication to high-volume or wide-variety production; it's rare to find a company that does both. In flexible printed wiring (FPW) the choices translate to a roll-to-roll or panel process. It's likely that most readers of this book will use panel process technique, or work with suppliers who do, but a few words on roll-to-roll technology at the outset would not be wasted insofar as they help clarify the managerial and technical issues. For the bulk of the chapter we'll concentrate on the more common FPW panel process and the unique manufacturing technology that's used there.

Manufacturing Overview

The raw material for FPW—the foils, films, and laminates—are manufactured by a roll-to-roll process. The same goals of tight control and low cost which operate at the materials level also apply to FPW: the ideal is roll-to-roll processing.

Because the necessary capital investment is enormous, there are only a handful of roll-to-roll FPW facilities in the world, most of which are dedicated to high-volume, low-cost circuitry used in camera, disk drive, automotive, and similar applications. These huge machines are capable of a narrow range of processes, typically producing single-sided FPW by the subtractive technique or screen-printed polymer thick film (PTF).

The roll-to-roll technique uses frequent, repeating tooling hole patterns (sprocket holes) for best registration; continuous operation produces huge volumes of circuitry in spite of narrow webs and low throughput speed. Conceptually, these lines operate as though they were running a continuous series of small panels joined end-to-end: as each pattern of tooling holes enters a process area, automated devices line up to them and perform an operation—image application, coverlayer, outline, test. Human operators are there to supervise, resupply, and remove completed rolls, but not to handle individual circuits.

Panel processing requires much lower equipment and capital investment, but consumes more labor per circuit. Increased labor and greater process diversity means more effort is required to keep the process under control.

Panel sizes range widely, depending on circuit size and required precision. Smaller panels aid alignment and therefore make more precise circuits, but larger panels and bigger tooling, if able to make acceptable product, result in lower cost. The smallest panel is a single circuit; largest common size in FPW is 18×24 in.

One way to compromise between large-panel/low-cost and single-circuit/high-precision strategies is to image and etch in large panels, then postetch-punch from etched fiduciaries to create tooling holes for each circuit which are used for subsequent coverlayer, throughhole, nomenclature, electrical test, and outline processes.

Calculating the cost to produce a circuit with the roll-to-roll process, which operates steadily, on the same design, for extended runs, is relatively easy. Working out the cost to produce in panels would be equally easy if one design was produced for a reasonable period of time—you'd simply divide the cost of keeping the shop doors open by the number of circuits produced. But the panel process is used by contract manufacturers, or job shops, who simultaneously run small quantities of different circuit types through different process sequences. In this situation, assigning equipment and overhead costs is an arbitrary process, at best.

Putting this all together, the distinctions are

- Roll to roll, for long runs without design or process change, is the least-expensive, highest-quality method. Amortization of capital investment and large initial tooling is offset by reduced labor.

- Panel process is best for short runs, complex and changing designs, or product development. Tooling changeover is less expensive and quicker; capital amortization is lower but labor content is higher; process control and product quality are issues.

Most FPW orders are for small volumes produced by the panel process; few reach the size where roll-to-roll manufacturing is justified. Panel production of FPW uses all sorts of chemical and physical technology in a range of process sequences which is difficult to set forth in an organized and easily understood manner: the enormous interactivity of materials and process defeats the best attempts at explaining production flow and technology. Varying process sequences even complicate the layout of panel facilities with the result that visitors will not find two plants with the same arrangement, and will find that a given plant has been rearranged by the next visit.

By either method, FPW is a challenging product to manufacture. No matter which method is used, there are general topics which should be discussed before we look at product-specific process sequences. See Chap. 3 for a discussion of differences between FPW and PWB production.

Materials

FPW processes are especially influenced by properties of raw materials and consumable supplies. It's a business where raw material properties impact yield, laminate choice dictates process sequence, and assembly technique determines surface finish. Therefore, incoming inspection and test to assure consistent performance is a critical part of the manufacturing process and is ignored at great peril: given the complexity of technology involved in FPW manufacturing, every possible source of variation should be pursued and eliminated for best yield.

The following material quality control tests should be performed, at least:

1. For laminates
    a. Thickness of dielectric, adhesive (if used) and foil
    b. Peel strength
    c. Dimensional stability
    d. Thermal stability

2. For coverlayer materials
    a. Adhesive thickness
    b. Volatile content
    c. Adhesion
    d. Flow

3. Foils
    a. Adherability
    b. Solderability
    c. Metallurgical: grain size, purity
    d. Smoothness/surface finish

Laminates

FPW laminates are produced by both adhesive bonding and by adhesiveless methods (see Chap. 7 for further discussion). Adhesive-bonded laminates consist of a dielectric film coated on one or both sides with a thin layer of adhesive by which they are bonded to metal foils. "Adhesiveless" materials—new to the FPW scene—are made by three methods: foils are coated with a polymer, polymer films are directly metallized by sputtering or chemical deposition followed by electrolytic plating, or films are bonded to foils by "high-performance" adhesives that approach polyimide in their properties.

Each type has strengths and weaknesses. Conventional adhesive-bonded products are lower in cost and have a long history; they're well-understood and accepted; adhesiveless materials are more expensive and unfamiliar but have superior thermal, chemical, and dimensional properties and can be significantly thinner.

Adhesive-bonded laminates are generally purchased but are sometimes made in house as a way to lower cost and utilize idle laminating equipment. In-house lamination is also used to produce custom laminates for the reverse-bare process by bonding together preperforated adhesive-coated films and foils (see "Lamination" and "Single-Sided FPW Processes" below for details).

No matter where it was made, a laminate must have solid process and materials pedigrees because circuit yield is a sensitive function of laminate properties. These complicated composites should be inspected to verify that film and foil type, thickness, and surface finish are as desired. Further inspections are required to determine dimensional stability (which can vary widely depending on residual film stresses) and bond strength, electrical properties, and resistance to thermal stress, all characteristics that are primarily determined by the adhesive.

Coverlayer and covercoat

The procedures for insulating a conductor pattern while leaving termination areas bare are unique to FPW and constitute one of the significant distinctions between FPW and PWB processing. The ideal technique is to use a dielectric which is identical in composition to the base dielectric because this yields technically and financially balanced and consistent properties—high-performance, high-cost base with matching coverlayer and so forth. Such a construction is equally good in withstanding environmental stresses on either side of the circuit, introduces the least variety of inventory items into the factory, and reduces the variety of chemical process effect. This is the rationale for *coverlayering*; in this process the insulating layer is composed of the same dielectric film and adhesive as the base laminate and is applied to the circuit in a lamination process which is similar (except for use of conforming press pads, as shown in Fig. 8-6) to the method for making the base laminate.

Covercoating is applying liquid or film dielectrics to the conductors by coating or low-pressure lamination followed, in some instances, by photodefinition of apertures at termination sites, and finally by crosslinking through thermal or ultraviolet (UV) cure.

Any coverlayer or covercoat process which involves heat—and most do—affects the dimensional stability of flexible laminates and thus the cost of FPW. This is a good reason for using covercoat processes based on UV or low-temperature cure.

Coverlayer process. The defining characteristic of the coverlayer technique is the method for creating apertures at termination sites. These may be formed either before or after the coverlayer is laminated to the circuit. If it is formed before lamination, adhesive flow into the apertures and registration or alignment are prime concerns. If it is formed after lamination, cost rises but aperture location and size are more precisely controlled.

Typical prelamination aperturing techniques are numerical-control (NC) drilling and punching. In both techniques the coverlayer film, with adhesive coating and protective release film (which guards the adhesive layer from handling damage and dirt), is pierced by a pattern of tooling holes and apertures which are sized slightly larger (0.005 to 0.01 in) to allow for adhesive flow. Punching produces the cleanest, least-disturbed holes but the tooling takes longer, and costs more to get, and has reduced changeability. Punching produces cleanest cuts when done from the adhesive side, a fact which influences tool design. See Figs. 8-1 and 8-2.

Drilling is cheaper for prototype runs and allows easily changed patterns but can damage the adhesive through drill heat unless process conditions (feed, speed, entry material) are carefully controlled (see Fig. 8-3). Both techniques are risky in terms of foreign material considerations.

However produced, apertures are of little value unless accurately registered to the etched pattern and kept free of adhesive fill during lamination; the combined effect of these factors must not reduce the annular ring or exposed area around a throughhole below prescribed limits (see Chap. 4, "Design"). Aligning coverlayers to the conductor pattern is a difficult task which requires skill and exposes the product to foreign material as the release film is removed and static charges build; it is best done under clean room conditions as shown in Fig. 8-5.

Aligning is accomplished either by pinning conductor and coverlayers together through common tooling holes or by operator eyeballing. Even when perfectly aligned and tacked in place, coverlayers may slip in lamination, producing rejectable product.

Coverlayer lamination. Coverlayer lamination requires enough heat and pressure to temporarily liquefy the adhesive, causing it to wet onto the etched pattern and top side of the base dielectric. Good lami-

130 Chapter Eight

Figure 8-1 A clicker press. An operator holds a steel rule die, used to cut rectangular pieces of covercoat material. Beneath the die is a tool-support drawer, which is pushed into the press after the material and die are aligned as desired. The operator presses two safety palm switches to close the press and cut the material. (*Courtesy Teledyne Electronic Technologies.*)

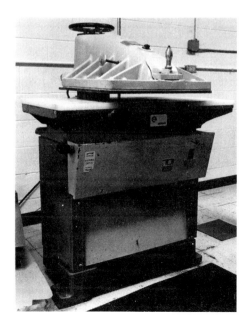

Figure 8-2 Another form of clicker press. The upper platen pivots around a vertical column and is swung to one side for alignment of tool and work on the bottom platen. The upper platen is swung back in position and forced downward by the press when the operator pushes the release button in the center of the handle. (*Courtesy Teledyne Electronic Technologies.*)

> 1. Stack up to eight sheets of coverlayer, adhesive side up. Interleaf the sheets with entry and backup material, such as aluminum foil or aluminum foil on chipboard. Total stack height should not exceed 10× drill diameter.
> 2. Clamp very tightly together (use full-area clamping to flatten the material and minimize wrinkles) and pin the coverlayer material and entry/backup stack together, adhesive surface upward (drill enters adhesive first; polyimide film rests on aluminum-faced backup material).
> 3. Set chipload at 2 mils/revolution [chipload = (infeed, in/min)/(drill speed, r/min)].
> 4. Set surface feet per minute (SFM) at 400 [SFM = (drill speed, r/min) × (drill circumference, in), where circumference = (π × diameter, in)/12].
> 5. Retract drill quickly; let it dwell briefly to cool before reentry.
> 6. Excessive infeed causes punchthrough and inadequate cooling; insufficient infeed aggravates the effects of spindle runout and dulls drills. Excessive stack height will cause overheating, burring of the release sheet, and "nailheading" of the coverlayer film. See Fig. 8-4.

Figure 8-3 Conditions for coverlayer drilling.

nation requires voidfree fill, which means plenty of flow; under these conditions the adhesive will also flow into the apertures unless care is taken to preclude this affect.

The quality of the lamination process is affected by:

Applied pressure

Timing of pressure application

Lamination temperature and rate of rise

Press pad

Adhesive thickness and volatile content

Conductor surface finish

Typical laminating pressures are in the area of 250 to 300 psi. Because coverlayer adhesives are designed for limited flow (to help keep apertures clear), full pressure is applied immediately with 10 to 20°C/min rate of rise to lamination temperature (180°C is common); this combination maximizes flow to improve fill around etched features and shortens cure time to restrict maximum flow. Excessively

Figure 8-4 A two-spindle drilling machine with noise-reducing cover open. The control/input station is at right, and 10 tool holders with drill bits are at the front edge of each spindle area. (*Courtesy Teledyne Electronic Technologies.*)

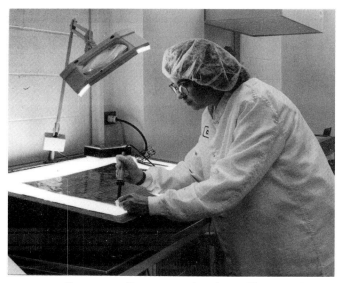

Figure 8-5 Covercoat alignment and tacking. The procedure is done in a clean room with smocks, hair covers, and laminar airflow from overhead. (*Courtesy Teledyne Electronic Technologies.*)

fast heat-up may cause premature sealing around the perimeters of the makeup, trapping volatiles. If excess flow occurs, reduced pressure or temperature or rate of rise may be employed.

Controlling adhesive flow is the prime function of the laminating press pad, specifically the release and hydraulic layers (see Fig. 8-6). With good technique, flow into apertures is controllable to 0.005 in or less. Press pad design is an art; there are many favorite systems. Common features are use of throw-away materials and combination of release film with hydraulic backup which readily conforms to any pattern design for best stoppage of adhesive flow.

Flow is strongly influenced by adhesive thickness, which is in turn driven by conductor thickness. Less adhesive is required for voidfree encapsulation of thinner conductors. Flow is essentially nonexistent

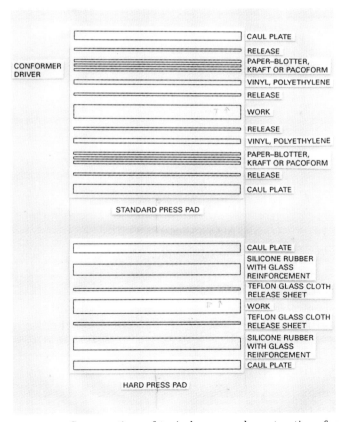

Figure 8-6 Cross sections of typical press pad constructions for bonding coverlayers or multilayer stacks.

Foil thickness, in	Coverlayer thickness, in	Adhesive thickness, in
< 0.0014	0.001	0.001
0.0014	0.001	0.001–0.002
0.0028	0.002	0.002–0.003
0.0042	0.003	0.003

Figure 8-7 Adhesive thicknesses for coverlayering.

with 0.001 in or thinner adhesive coatings, which are adequate for conductors 0.001 in thick and for PTF.

Proper adhesive thickness is determined by adhesive properties, foil and coverlayer film thickness, and pattern density, with more adhesive needed to fill around etched patterns in thicker foils, where center-to-center distance between runs is reduced or where greater film thickness reduces conformation. The least thickness that fills adequately is best; it's typical to use adhesive thicknesses roughly equivalent to the foil thickness, somewhat more for dense patterns, with thicker coverlayer films or where flow into apertures isn't a major problem (Fig. 8-7).

Large apertures are more easily protected from flow because the press pad can get down into them more easily and therefore can tolerate greater adhesive thickness. Smaller apertures—less than 0.06 in in diameter—are more difficult to protect and are proportionately more influenced by small (0.005-in) adhesive flow, and thus require reduced adhesive thicknesses.

Inadequate adhesive flow, which can result from deviations in any of the listed parameters, leads to inadequate sealing which appears as air bubbles or "soda straws" along the sidewalls of etched runs. This phenomenon is detected by refraction of light from the air-adhesive interface and is one of many examples of the impact of transparency on FPW manufacture (see "Comparison with PWB Process" in Chap. 3). An easy inspection for soda straws involves shining a sharply defined beam of light along the conductor edges and looking for refraction. Another method is to cut a sample with several parallel conductors to expose the coverlayer/conductor interfaces, then expose the cut surface to a saturated solution of sodium sulfide for 30 minutes, allow the surface to stand at room temperature for 24 hours, and examine it for black stains.

Surface preparation. Conductor surface preparation for a coverlayer is a sadly ignored issue in FPW production. Oxide treatment—the production of stable, chemically inert surfaces for enhanced adhesion and storage stability—is invariably used in PWB manufacture but unheard of in FPW. A possible reason may be the expectation that alkaline solutions will attack and degrade FPW dielectrics. Experience indicates that time, temperature, and solution concentration are factors in both oxide treatment and dielectric attack and that there's a large process window through which adequate surface passivation can be achieved without damage to the dielectric.

Common FPW production practice is to prepare for coverlayering by cleaning the conductor surfaces and storing under "tarnish resistant" conditions to await lamination. Copper foil treated in this fashion develops a poorly adherent, easily attacked layer of cuprous oxide. The absence of a chemically stable, adherable conductor surface leads to reduced coverlay adhesion and a phenomenon experienced during solder exposure called *solder wicking*. Wicking is frequently attributed to coverlayer delamination but actually occurs when solder flux, whose purpose is to strip off oxides and allow molten solder to wet onto the base metal, attacks and strips away the weak natural oxide, allowing solder to penetrate beneath the coverlayer.

Drilled coverlayers show this defect to a greater degree than punched coverlayers, suggesting that drill heat plays a part. Modified epoxy coverlayer adhesives are better at resisting solder wicking; acrylic-based systems have poor resistance.

Other methods aimed at improved coverlayer adhesion and resistance to solder wicking are use of *double-treated* foils (both sides are given an adhesion-promotion treatment by the foil manufacturer; see Chap. 6, "Conductive Materials") or any of a wide range of proprietary surface-treatments.

Postlamination aperturing. Several techniques are employed to create apertures—a process called *pad baring*—after coverlayer lamination. Baring an already coverlayered FPW is difficult because the coverlayer must be removed without damage to the delicate foil termination. Advantages of postlamination baring are that the openings can be accurately located according to drawing coordinates and they can be precisely defined because adhesive flow isn't a factor. However, cost is high because of the required skills and poor yield, and location to coordinates is only part of the aperture problem—openings must also align with the etched pads. If we create apertures correctly located 10 in apart, but which fail to align to etched pads that are *not* 10 in apart (because of postetch shrinkage and other instabilities), we've not improved our yield. This dual-dimension dilemma is a common problem in FPW manufacture and has no easy solution so long as customer designs contain both artwork generated by computer-aided design (CAD) and dimensional drawings.

Methods for postcoverlayer aperturing are

Machining—skiving, spot facing

Laser ablation

Chemical etching

Mechanical machining is only practical where a few apertures are required, because this is a one-at-a-time, high-cost, high-skill operation with great risk of pad damage. Chemical etching or laser ablation are safer processes which remove the coverlayer with less risk to the pads by taking advantage of differences in energy absorption and/or composition. The cost of chemical etching isn't related to the number of openings, while laser ablation and machining both cost by the opening; if only a few openings are required, the expense of mechanical or laser baring may be offset by avoidance of the cost of preperforation and register lamination.

Openings in FPW are ordinarily in the coverlayer but may be in the base layer; when openings are in both base and coverlayer the design is said to be *double-side bared*. This is an instance where custom laminate production and postlamination baring may be cost-effective.

FPW intended for PTH termination doesn't require apertures and can be coverlayed with high flow adhesives and hard press pads which yield a solidly filled, flat and uniform circuit. From this point of view, plated throughhole- terminated FPW represents a much simplified materials and process control problem. (See "Plated Throughhole Processes," later in this chapter and Chap. 9, "Rigid-Flex Manufacture.")

Covercoat process

Insulating the etched pattern by application of a photoimageable liquid or film coating eliminates the expense and labor of preparing and registering a coverlayer and the lamination process. Performance of covercoats is lower than provided by coverlayers because cured-in-place covercoats can't be separately stretched or oriented for maximum mechanical properties, can't be cured or cross-linked for best properties without subjecting the FPW to potentially destabilizing conditions, and are generated by a compromise of coating properties, curing reaction, and coating thickness. Dielectric films used in coverlayer process can be produced by whatever unique methods are required for best performance, then bonded onto the etched pattern by a polymer chosen specifically for adhesion, activated through high lamination temperatures and pressures. Liquid and photodefined covercoats must do all this from one chemistry, at one time, on the FPW.

Covercoats can't easily achieve high flexibility—defined as crease resistance—required for FPW use together with enough cross-link density for adequate thermal and chemical resistance. Both liquid coat and photodefined film techniques are used in PWB production, but in PWB use there's no need to compromise cross-link density for better flexibility since it's not required.

Surface Finish

Copper is a reactive metal; something must be applied to bare terminations to preserve solderability or to allow pressure connection after storage. The FPW manufacturer provides this treatment; options are

- Noble metal platings
- Hot-air solder leveling (HASL)
- Organic protective coatings
- Electroless tin

Gold is used for PWB edge connectors and is sometimes found in high-reliability FPW designs. This finish is normally applied by electrolytic technique, usually over a nickel underplate, which means at a point in the process where the conductor pattern still has interconnections or "busbars."

It's good technique to plate after the coverlayer (or covercoat) is applied, because this guarantees alignment between coverlayer apertures and plated finishes, leaving on the one hand no exposed conductor surfaces and, on the other, no adhesive on the plating. To make it possible to electroplate after the coverlayer is applied, a busbar is included in the design which either extends past the outline to be cut off at final trim or buses internally and is disconnected by piercing or drilling through the finished circuit.

New developments in the technology of electroless deposition may make it possible to plate the exposed terminal areas of coverlayer circuitry to adequate thickness without busbar interconnection.

Solder coating, or tinning, gives excellent all-around protection for FPW terminals. It works directly with most subsequent (soldering) assembly techniques and is also used with many pressure connectors. Solder is a common PWB surface finish and could be used easily in FPW except that it's poor practice to have solder under the coverlayer or covercoat where it may remelt during assembly soldering, destroying the coverlayer or covercoat bond and opening a fissure into which flux will creep for later mischief, as occurs in solder wicking.

Thus solder must be applied after the coverlayering process, which presents the same bus problem—if the solder is electroplated—as does noble metal finish. Hot-air solder leveling is the answer to this problem. It's widely used and produces a low-cost, definably solderable pad surface, but the temperatures and forces required—prebaking to remove moisture, exposure to the solder wave, and subsequent air blasting to remove excess—are a threat to FPW and result in some degree of shrinkage, distortion, and delamination. However, if a solder finish is required, this is one of the best choices.

Organic protective coatings are fast gaining popularity. They're easily applied by the FPW manufacturer, don't abuse the product, and provide adequate storage stability. Systems are available from all major electrochemical suppliers; required exposures vary, depending on desired storage life and bath temperature, from 1 to 5 minutes at room temperature to 120–130°F.

Electroless tin deposition is another option, adaptable to all forms of FPW, easy to control, and harmless to the circuit. Assuming the pad is clean when the tin is applied, and given benign storage, this is a reasonable short-term method for preserving solderability.

Lamination

In roll-to-roll processing

As the section "Process Sequences" shows, lamination (assembling materials together in layers or, if you prefer, assembling layers of materials together) is frequently used in FPW production. In roll-to-roll manufacture it's accomplished by a set of nip rolls, one roll heated and usually metal, the other compliant (silicone rubber, because of its heat resistance, is a common choice) and forced against the heated roll under adjustable pressure. The webs feed through and are exposed to brief temperature and pressure. Exposure to heat is varied by changing the angle of wrap of material around the heated roll, adjusting its temperature, or changing web speed.

Roll lamination works well with thermoplastic adhesives and some thermosets, but duration of pressure application is too short to adequately bond many FPW adhesives. The method can be used to apply coverlayers, but it requires careful control of alignment and informed choice of adhesive and compliant roll materials to minimize flow. An upside-down technique is commonly used in the roll-to-roll process—see "Single-Sided FPW Processes" for details. In this method the base laminate consists of foil bonded to its coverlayer; the etch image is a mirror of the conductor image and the "coverlayer" step applies what is normally the base dielectric—no apertures.

Shuttle presses can be used in the roll-to-roll process: these are moving lamination presses which grasp the web, apply heat and pressure while traveling with it for a prescribed time, and then release and travel rapidly upstream to grab again. Dwell times can be extended by this technique, but to a limited degree—if the web moves at 3 ft/min, a 9-ft shuttle provides only 3 minutes of lamination. But fast-curing adhesives can be worked enough, in this short time, for satisfactory coverlayer adhesion.

In panel processing

Press lamination offers extended cycle times at widely varied pressures and temperatures from a low of 100 psi to 500 psi and more and at temperatures up to but not much more than 400°F, and is the panel-process choice.

Press loads or "books" are assembled, or *made up,* in clean-room conditions to minimize foreign material. Books consist of caul plates, which protect press pads and work from handling and contact with somewhat rough and dirty press platens, inside of which is a press pad on each side of the *work,* or FPW, to be laminated. See Fig. 8-6 for standard press pad constructions.

Press pads provide three functions:

1. Absorb variations in platen and load thickness (conformation)
2. Protect apertures from adhesive flow and level out pressure by means of a semimelting, hydraulic layer (driver)
3. Protect the work (release) from adhesion to the hydraulic layer

It's important to note that laminating press platens are always bowed to some degree by constant temperature cycling; caul plates are never flat and books always vary in thickness. Engineers calculate applied pressure by dividing the applied force by the platen area—22 tons (44,000 pounds) applied to an 18- ×24-in platen represents about 100 psi *average*. But since press loads aren't uniformly thick over their full area and the press pad has inadequate conformation ability, average pressure won't be relevant, because all the press force will concentrate on the thickest area, applying high pressure there and correspondingly reduced pressure in thin areas. This is a very complex issue which is addressed by use of the conforming layer, whose purpose is to crush down in the high areas, thereby transferring force (and pressure) to low areas. Conforming and hydraulic layers help, but careful attention to flatness and uniformity is necessary. Note that any tendency for hydraulic or conforming layers to spread laterally results in distortion of the FPW.

Typical conforming pads are made of steel wool, multiple plies of crushable paper, or fibrous mats (Fig. 8-8). Steel wool is reusable but coarse, tending to imprint the work; papers are popular and available

Conformer	Advantages	Disadvantages
Kraft paper	Low cost	Uneven
Chip board	Low cost	Uneven
Silicone rubber	Uniform, durable	High cost
Ceramic-fiber felt	Excellent uniformity and conformation	High cost, single use, messy

Figure 8-8 Chart of press pad materials.

in a range of uniformities, crushabilities, and costs. Fibrous mats are the most expensive but also provide the best compensation for unevenness without lateral distortion.

Hydraulic layers—which are always thrown away after use because they are imprinted by flow into the apertures, tooling holes, and so forth—are composed of polyethylene in various densities including radiation cross-linked, vinyls, and proprietary materials such as PacoForm™ and PacoThane Plus™ (Paper Corporation of America) and PAL (Gila River Products Inc.). Release films can be Tedlar,* Teflon FEP,* or skived TFE; TEF-Teflon–impregnated glass cloth; or proprietary systems consisting of combined release and hydraulic layers. All release layers have conflicting requirements: they must move in the Z direction and imprint into the finest details of the FPW surface, to inhibit adhesive flow, but they must not allow lateral movement (in the X-Y plane) of the hydraulic layer to stretch or distort the FPW.

Platen loads can be heavy enough for an operator to require assistance in loading and unloading. An arrangement of roller conveyer line with elevator section is used to move loads from the layup room to the press and back again for breakdown after lamination as shown in Figs. 8-9 and 8-10.

Desiccation

Most materials, particularly polyimide films and acrylic adhesives, absorb significant amounts of moisture, even under air-conditioned factory conditions. Moisture or other volatiles in a press makeup cannot readily escape through the multilayer makeup at lamination pressures. Therefore, desiccation is a very important manufacturing consideration. Baking is a low-cost method used on nonreactive materials such as etched details and stiffeners, but coverlayers, adhesive, and prepreg layers must be protected from elevated temperatures; desiccation of these reactive materials requires storage under dry gas (nitrogen) or vacuum conditions for extended periods (24 hours).

---

*Tedlar and Teflon are registered trademarks of DuPont.

Figure 8-9 Six-opening laminating press. Overall height is 6 ft. The conveyer line has an elevator section for handling books of material. Foot pedals to control the elevator are at bottom right. (*Courtesy Teledyne Electronic Technologies.*)

Vacuum-assisted lamination

One common method of vacuum lamination, called *turkey-bagging,* consists of sliding the platen load into a temperature-resistant plastic bag which is connected to a vacuum system. Desiccation is provided by holding the load under vacuum for an hour or two prior to lamination to remove residual moisture and other volatiles; the bag remains connected to the vacuum during the lamination cycle and is cut off and discarded at break down.

Vacuum-assisted lamination is not commonly used for single- and double-sided FPW but is frequently used in multilayer and rigid-flex constructions which are very susceptible to lamination voids caused by volatiles.

Bagging is an expensive process, consuming roughly 20 minutes in labor to put on and take off each bag. This cost is avoided, and even better volatile extraction is enjoyed, by the use of vacuum presses as shown in Fig. 8-10. This six-opening press is identical to the press in Fig. 8-9, except that the platen structure is enclosed in a vacuum chamber. Books are loaded onto the platens, the door is closed, and vacuum is applied and maintained for the full cycle of prelamination devolatilation, heat-up, cure, and cool down. Because the loads are

142   Chapter Eight

Figure 8-10 Multiple-opening vacuum press. This equipment is similar to the press in Fig. 8-9; however, the platens are housed in a vacuum chamber. Automated process controls for vacuum, dwell, temperature, pressure, and cooldown are at right. The chamber door is open (to the left) for access. Elevator and conveyer system handles heavy loads. (*Courtesy Teledyne Electronic Technologies.*)

fully exposed along their perimeters (in contrast to being constricted into a bleed film in a turkey bag), the extraction is unimpeded.

Autoclaving

Autoclaving is a lamination technique which enjoyed brief and undeserved popularity in the '80s. This technique uses the same labor-intensive and expensive turkey-bag process found in vacuum-assisted lamination, but laminates by way of pressurized, heated gas in a pressure chamber. There are two advantages to autoclaving: it heats evenly (by means of the circulating gas) and it is insensitive to platen flatness because there are no platens. However, it's incorrect to call this *isostatic* pressing, as was the habit of autoclave salesmen, because it's necessary to have a rigid structure inside the turkey bags to support the tooling pins and keep the load flat as it heats; if this structure is at all uneven or bent—and all are—then pressure application is also uneven.

Disadvantages of autoclaving include:

High equipment cost

Dangerous pressures and enormous stored energy

Slow cycles—slow to heat, slow to cool

Limited maximum pressure: 300 psi

Turkey bag labor and material consumption

Imaging

Imaging is the process of transferring a master artwork onto a laminate in a form which controls or "resists" another process step to define the conductor pattern. The result of imaging is called an image; images are defined as *subtractive* or *additive,* and can be *positive* or *negative* with respect to the art master.

The two principal imaging media in FPW production are silk screen and photoresist; silk-screening literally prints the image directly onto the laminate in ink or paint, while the photoresist process requires three steps: resist coating, exposure, and development.

Images for subtractive processing control etching; the conductor pattern is protected by the image and remains after unresisted areas of the foil are removed. Additive images protect the opposite areas— the space between conductors. They're used to control electro- or electroless deposition: conductors are built up (added) in the image, defined by its sidewalls up to the thickness of the resist. The additive technique is the more precise method for creating thicker patterns— above 0.001 in—because it's degraded only by imaging errors, while the subtractive process has two variables: image error plus the effect of etch undercut or *etch factor.* In thin conductor patterns, where undercut is negligible, the difference between the two techniques disappears.

Silk-screening

Silk-screen printing is a technique which is borrowed from the art world. It's a method widely used to inexpensively print placards and small signs, tee shirts, and posters. Screening consists of forcing a thick, highly thixotropic resist through a patterned, fine mesh screen (110 to 300 wires per inch in each direction) and onto the panel, which is held by a vacuum chuck. It applies a very thick and uniform coating: in artistic use, it builds bright, vivid colors; in PWB/FPW production, sturdy, etchant-resistant images with resolution that is rheology-limited to about 0.008-in lines and spaces. Screen printing can't be easily done on perforated panels, and won't tent over holes: you can screen print only onto something which is there.

Positive and negative images

An image is said to be *positive* when the resist pattern looks the same as the art master; i.e., the resist is present where the artwork is opaque. Negative-working photoresists produce an image that's opposite, or *negative,* compared to the artwork: where the artwork is clear, the image is present. Thus, for a subtractive (etching) process with negative-working resist, areas in the artwork which represent the conductors will be clear. Similarly, in an additive process where conductor areas are exposed in the resist pattern, conductors in the artwork are represented as black areas. Figure 8-11 shows the relationships.

Photoresist application

A major process consideration is whether the resist is applied as a liquid or dry film. Earliest resists were inks or paints which were stenciled or silk-screened in a desired pattern onto areas to be protected. Development of photosensitive liquid resists offered a better production technique in which images of a master pattern could be replicated with greater precision and without the bother and mess of making a master screen and dealing with large quantities of paint. But liquid resists produce thin coatings whose protection is easily violated by tiny particles of foreign material; use of liquid resist therefore requires extremely clean, smooth laminates and clean coating and drying conditions to produce sound images.

To get around these requirements, the so-called dry-film resist process was developed in the late '60s. In this technique the photosensitive resist coating is applied to a temporary carrier and dried to form a film under clean-room conditions. The resist film is later transferred from the carrier onto an FPW substrate to yield a relatively thick photosensitive coating.

Because they're thicker, dry film images are more tolerant of dirt and foreign material, are tough enough to survive indifferent care in etching or plating, and allow additive production of conductors up to thickness of the film—anywhere from 0.001 to 0.004 in as compared to 0.0002 to 0.0004 in in liquid resist images.

Process type	Resist choice	Artwork conductors are
Subtractive	Negative	Clear
Subtractive	Positive	Black
Additive	Negative	Black
Additive	Positive	Clear

Figure 8-11 Relationships of process type, resist type, and artwork opacity.

Dry film is transferred to the substrate under moderate heat and pressure, by either a continuous hot-roll or cut-sheet technique. Cut-sheet equipment creates a resist coating which is smaller than the panel size to provide a peel-resistant outer edge; roll lamination to FPW panels requires constant operator attention and careful cutting between panels. See Figs. 8-12 and 8-13 for dry film application equipment.

Dry film will *tent,* or bridge, over holes to protect them from etching, and for most FPW manufacturers, it's the preferred imaging medium because it tolerates minor process variations and dirt and has adequate resolution for current and near-term future designs. Photoresist—dry film and liquid—is available in both negative- and positive-working systems, which helps to simplify artwork inventory.

Liquid resist is applied by roller coating (*gravure,* one or two sides) or by dipping and controlled withdrawal; solids content and roller or withdrawal speed control thickness. Recently, manufacturers of liquid resists have challenged the dominance of dry film with electrophoretic application. This is a method that resembles electroplating: the part to be coated is immersed into an emulsion of resist, then electrified so that it electrostatically attracts the resist to itself in a uniform film. This technique allows uniform coating into and around holes—difficult with direct wet application—and buildup of very thick (for liquid resist) coatings of up to 0.001 in.

Figure 8-12 A hot-roll laminator for dry film. The machine operates at a temperature of 223 to 244°F, a pressure of 2 to 5 bars, and a speed of 6 to 10 ft/min.

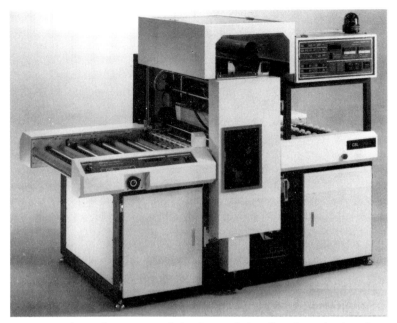

Figure 8-13 A cut-sheet automatic laminator. It handles sheets in sizes from 10 × 13 in to 24 × 25 in at a rate of 220 per hour.

Exposure

Once the resist coating is in place and sufficiently dried (or cooled, in the case of dry film; 15 minutes cool-down, or *holding time,* after application is typical), it is exposed to a strong UV light source through the artwork or photomaster. This process takes place under semi-clean-room conditions to minimize foreign material (particularly important with negative-working resists, which are removed in areas where exposure is blocked by dirt) and requires tight contact between artwork and resist for maximum image quality.

Unified exposure equipment provides collimated (parallel-ray) light and vacuum contact. Absorption of heat is another function of this equipment—UV light sources produce plenty of infrared as well, which must be absorbed and shunted into water or air cooling circuits to protect the artwork from excessive heat. Heat is not important for one or two exposures, but after an hour or so of production use, sufficient heat can accumulate to both expand artwork out of tolerance and interfere with resist exposure.

Pull-out drawer systems are found in much panel-process manual equipment; artwork is pinned to one or both sides of the panel, then the assembly is dropped into the drawer. The bottom of the drawer is glass, and the hinged top is a sheet of plastic film which seals around the edges and is drawn down onto the artwork, panel, and pins by an internal vacuum system to force intimate contact. Where glass artwork is used, the same procedure is followed with compensation for the increased thickness.

Older exposure systems split the light into two beams to simultaneously expose through the glass and plastic surfaces. The latest approach is to concentrate the energy on one side at a time, because shorter exposure is better. The energy requirement for most resists is about 300 mJ/cm^2.

Automated systems in roll-to-roll processes or highly automated panel factories align artwork by means of cameras and fiduciary marks, eliminating tooling holes and the inevitable error which creeps into pinning. Use of three or four cameras and polling software reduces alignment error to 0.001 in or less. Full automation prolongs artwork life and reduces foreign-material problems because the equipment is sealed—no human intervention, no human debris.

Collimation is a central theme in discussions of fine-line reproduction but is generally overrated in importance. With thin resists (less than 0.001 in) and good artwork contact, collimation isn't important until feature size drops below 0.002 in.

Development

So-called aqueous processing is standard in the printed wiring industry. A common developer solution is 1% sodium carbonate in water at 90°F. Equipment for development, etch, and strip processes is remarkably similar, consisting of a conveyer system, one or more spray chambers where process chemistry is powerfully applied to both surfaces of the laminate, and finally rinse and dry modules. Figure 8-14 shows a resist developer machine.

After development, some resists require a preetch bake to increase film strength and adhesion.

Photoresist warnings

- Because photoresists are formulated for rapid exposure, they must be handled under controlled light conditions; a typical arrangement is "gold"-colored fluorescents.
- Dry film has limited ability to conform to small dents and surface roughness, consequently production of fine lines (less than 0.005-in features) demands smooth foil surfaces, properly cleaned and prepared for adhesion.

Figure 8-14 Conveyerized developer with controls. (*Courtesy Teledyne Electronic Technologies.*)

- Lamination temperature must be controlled—high enough to get good adhesion, low enough to avoid stripping difficulty. Temperatures should be kept below 200°F; 180°F is better.
- Development must be continued until unexposed resist in the corners—the so-called knees—is adequately removed, a condition which is difficult to achieve in thicker resists with fine features.
- Choice of resist must take into account the pH of the etchant or plating baths which will be used. Most resists will withstand acidic conditions better than alkaline: because dry film strips in mild caustics, some versions are not recommended for use in ammoniacal etchant. For FPW with PTH termination, the common process is to use dry film resist in a pattern plating process that finishes with tin-lead plate, after which the dry film is stripped off and the tin-lead, or *metal,* resist protects the panel through an ammoniacal etch cycle. This method reliably produces PTH barrels and small annular lands which are covered with tin-lead overplate, which is a common requirement. Cupric chloride is acidic and thus doesn't threaten dry film; it is used for inner layers with dry film or liquid resist in a subtractive process.

- Etching of fine-line features is inhibited by thick resist patterns, which interfere with free solution flow into narrow spaces. A rule of thumb: total foil plus resist thickness determines minimum feature size; 0.0014-in foil with 0.001-in resist *may* produce 0.0024-in features, assuming excellent process controls and care.

Resolution

The ability to replicate fine features is called *resolution* or *resolving power*. There is a close correspondence between resist thickness and resolving power: thinner resists have higher resolution. Liquid resist systems which produce coatings 200-μin (0.0002-in) thick can resolve features which are several times finer than the thinnest available dry films (0.0007 in) and are therefore the choice for fine-line circuitry. Because they're thinner, images generated with the liquid resist process also interfere less with solution dynamics in the etch process, aiding production of fine spaces.

Dry film carriers must be left in place during exposure, because the resist is tacky; electrophoretic resist coatings may also require a release sheet between resist coating and artwork. These carriers are undesirable because added thickness spaces the artmaster further away from the resist, thereby degrading resolution.

In additive processing, the resist should be at least as thick as the pattern to control lateral expansion, or *mushrooming*; if the design requires conductors thicker than 0.0002 to 0.0003 in, dry film or electrophoretic coating is the best choice, but feature size will be limited to 0.002 in or larger. Plating must not be allowed to build thicker than the resist. Once the deposition reaches the surface of the image it will rapidly mushroom out over the resist surface, degrading the image and inhibiting clean stripping and thus potentially causing short circuits.

Tenting, nonflat images

A major advantage of dry film resists is that they can bridge over and protect (tent) holes, yet will also cleanly strip from perforated panels after etching. The disadvantage of this uniform thickness and low adhesion is that it's difficult to get a tight seal in applications where the laminate surface has step changes in height. Examples are dents or weave patterns in the surface or at pad edges in the selective plate process for FPW with PTH (see Fig. 8-39). The degree of difficulty is related to the ratio between resist film thickness and step height. Liquid resists easily coat and protect such steps. Silk-screen printing onto irregular surfaces is messy to impossible, depending on slope and magnitude of height changes.

Other imaging techniques

Electroforming. Electroforming is a process of creating objects (or conductor patterns) by electroplating onto a master of the desired shape. It's used to produce very complex shapes and/or highly polished surfaces in configurations or metals which are hard to fabricate by conventional means. A modified version of this technique has been used to produce large volumes of FPW. The process:

1. A semipermanent image is applied to a polished stainless-steel belt.
2. The belt passes through a continuous electrodeposition process which builds 0.001 to 0.002 in of copper in the exposed area of the pattern.
3. The plated belt then passes through a nip roll where an adhesive-coated film is bonded over the pattern.
4. The film with conductors is stripped off the belt (copper doesn't adhere strongly to polished stainless).
5. The belt is replated for another pass, and so forth.

The advantages of electroforming are low cost, exact image replication, freedom from handling damage, reduced shrinkage, direct appli-

Figure 8-15 An electroformed circuit.

cation of adhesion-promoting surfaces, ability to produce very thin conductors and/or precious metal finish on contact areas, and, by use of preperforated dielectrics, two-sided access without added labor. See Fig. 8-15 for a sample circuit.

PTF or conductive ink. This is not an imaging technique in the sense that it regulates or controls another process; in the PTF process, the image *is* the conductor. PTF takes advantage of the ability of screen printing to apply a relatively thick mix of polymeric binder and conductive particles to a substrate, after which the binder is dried or cross-linked to create a solid matrix of conductive particles in the desired conductor pattern. Because it eliminates many wet process steps and environmental issues, this method has potential significance for FPW production. However, compared to copper conductors, PTF patterns have severely reduced conductivity (3 to 4% of copper), very low current-carrying ability, and are not solderable, forcing use of nontraditional termination schemes such as conductive adhesives, low insertion force (LIF) connectors, and the like.

Another issue is toughness, that is, crease resistance and flexural endurance. To get even modest conductivity, loadings of conductive particles in the ink must be high (70 to 80%), which means the cured PTF pattern has poor mechanical properties. Nevertheless, this method is attractive for use in cost-sensitive, benign applications where conductive adhesive assembly is acceptable. With time, reliability concerns will be resolved.

Die stamping. This is also not a true imaging process, because die stamping produces the finished conductor pattern in one step. For further discussion, see "Alternative Imaging Processes," below.

Imaging summary

Screen printing is employed for high-volume, relatively coarse designs produced by subtractive technique.

Liquid resists are used for finest features (less than 0.003 in) and on irregular surfaces which would be difficult to protect with dry film. Electrophoretic application provides liquid resist performance in thicknesses to 0.001 in for additive processing.

Dry film is a tolerant photoimaging process which is used for subtractive, PTH, and additive processes where feature size is 0.003 in or greater.

Electroforming can be employed for high-volume production where electrodeposited conductors are acceptable.

The PTF technique is applicable to low-current, low-conductivity interconnections and benign environments where it offers reduced cost and process simplicity.

Etching

One of the most heavily engineered and regulated processes in FPW and PWB production, etching is an ancient metalworking technique which has continually improved as designs trend ever smaller. Etching attracts disproportionate attention not only because it's basic to production and product quality, but also because, as a wet process, it's the focus of environmental concerns.

The speed and quality of etching depends on:

1. Temperature
2. Agitation
3. Chemistry

Equipment

Temperature and agitation are controlled by production machinery; see Figs. 8-16 and 8-17 for an overall view of one example. Most etchers are conveyerized and have one or more spray sections with independent controls, oscillating spray bars, top-and-bottom adjustments, and a sump into which the spent etchant falls to be replenished or replaced. Following stations or compartments along the conveyer

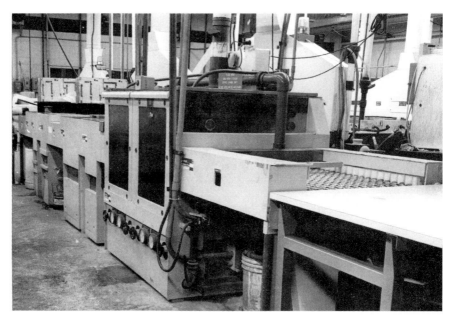

Figure 8-16 Conveyerized etcher, left side view. (*Courtesy Teledyne Electronic Technologies.*)

Figure 8-17 Conveyerized etcher, right side view. (*Courtesy Teledyne Electronic Technologies.*)

rinse and/or neutralize etchant residues, and apply postetch chemical treatments—for example, tin-lead brighteners—if required.

Spraying is employed to aid in forcing fresh etch solution into tight spaces, thereby increasing production speed. Spraying imparts a degree of directionality to the process, which tends to increase the etch rate in the direction of spray momentum (into the foil) and to relatively reduce undesired lateral etch, which reduces line width.

Etchers are built in both horizontal and vertical orientations. The claimed advantage of vertical etchers is that the process is more uniform because there's no puddling effect, as found in horizontal machines, to inhibit impingement and thus upset uniformity. However, there is a tendency for greater etching on upward edges because the solution always runs off in one direction.

Horizontal equipment is simpler to build and maintain and is more common. As panel size increases, pools of etchant collect on the top surfaces and slow down the action, forcing adjustment of top-to-bottom nozzle pressures or use of only the bottom—which doesn't suffer from puddling—for delicate work. Shutting off the topside nozzles results in halving the throughput of double-sided panels but increased control may offset added cost in delicate, fine-line (less than 0.005 in) work.

The spray pattern and operating pressure have enormous impact on etch speed and uniformity. Each machine must be carefully evaluated by measurement of test panels and adjustments to throughput speed, pressure, nozzle size, oscillation rate, and chemistry to get the best compromise condition.

Conveyers, as noted in Chap. 3, have trouble handling FPW laminate. As shown in Fig. 8-18, a leader may be taped to FPW laminate to guide it through a conveyer system. Details of a conveyer design are shown in Fig. 8-19; here the top and bottom roller bars are seen with drive gears at each end, together with spray rinse bars. The purpose of this complex design is to provide maximum exposure of both top and bottom surfaces of the laminate to the process while reliably transporting it. Topside rollers are needed to hold lightweight FPW materials in place against forceful sprays, but the required close spacing of roller bars and wheels interferes with spray uniformity.

Very fine line work is done in planetary equipment where the work (FPW laminate must be racked) is fastened to a revolving vertical table and exposed to spray nozzles which may be either fixed or oscillating. Rotating the work eliminates sidewall effects while nozzle oscillation smooths the pattern.

Still or immersion etching is slower but sometimes practiced, particularly for very thin metals (less than 0.0005 in) where exposure time is

Figure 8-18 Detail of a leader taped to a flexible sheet. (*Courtesy Teledyne Electronic Technologies.*)

Figure 8-19 Detail of a conveyer, showing top and bottom rollers and spray bars. (*Courtesy Teledyne Electronic Technologies.*)

measured in seconds. In this method the FPW panels are racked about 1 in apart and inserted vertically into an agitated bath. Immersion etching requires very simple equipment and controls; assuming good choice of chemistry, it can produce excellent etch quality.

Chemistry

Choice of etchant affects cost, speed, and precision. Since large solution volumes are involved, an overriding factor in etchant choice is the question of what to do with the spent solution. Some etchant manufacturers haul the waste away and regenerate it; this attractively simple solution is used by most FPW producers. Because these are proprietary systems (based on ammoniated cupric chloride chemistry), chemical cost is high but overall cost, including disposal, is competitive.

In-house regeneration is accomplished by extracting excess copper through electrowinning or plating out or by crystallization. Regeneration has the attraction of closed-loop purity but requires sizable investment in equipment and skilled staff to keep the system in balance. It keeps the solution at its intended strength for a consistent process. Another technique is replenishment, in which a portion is bled off and replaced by fresh solution. However accomplished—by replenishment or regeneration—it's critical that etching solutions operate at predictable strength, because solution activity affects conductor width.

All commercial, high-volume etching solutions are designed for copper, because copper is far and away the most commonly used metal in FPW and PWB production. Where special metals are to be etched, different etchants such as ferric chloride are required. Specialty etchants, because they are used in low volumes and are difficult to regenerate, are expensive both to buy and to get rid of; ferric chloride costs more to purchase than cupric chloride and double that for disposal! Figure 8-20 compares characteristics of etchants.

Ammoniated CuCl is preferred for production of PTH with metal-plated (tin-lead) resist but can be difficult to operate with dry film resists, which are sensitive to attack at high pH values. Ammoniacal replenisher solution is normally added at the flood-rinse stage after the etch chamber, where its high-ammonia, low-copper chemistry aids in removing surface stains or ionic contamination. However, if the pH is too high at this point, because of heavy replenisher flow, the dry film image may be damaged and not survive a second, or cleanup pass if one is needed.

Cupric chloride is the cheapest system to operate in terms of dollars per dissolved ounce. It has about half the speed of alkaline etchant, which could force purchase of additional equipment and requires careful management of replenishing chemicals such as hydrogen peroxide and hydrochloric acid. The inherent etch factor of cupric chloride is very good.

Hydrogen peroxide–sulfuric acid systems are not popular but offer closed-loop operation by refrigeration/crystallization of copper sulfate. It is a clean-working bath with low environmental impact but expensive in chemistry and management.

	Cost per ounce, cents	Speed, mils/min	Etch factor*	Regeneration
Ammoniated copper chloride	8–12	2.5–3	0.2–0.4	Yes
Cupric chloride	4–8	1	0.4–0.6	Yes
Hydrogen peroxide/ Sulfuric acid	15–50	1	0.4–0.6	Yes
Ferric chloride	>50	0.5–1	0.4	No

*In copper, with banking agent

Figure 8-20 Comparison of FPW etching solutions.

Etch factor

An important etchant property is the etch factor, or degree of undercut. Undercut (under the resist) results in conductors which are narrower than the artwork. There are several conventions for expressing the etch factor; we use the ratio of narrowing *per side* divided by thickness. Thus, if a conductor measures 0.01 in at the bottom and 0.008 in at the top and is 0.0014 in thick, the narrowing is 0.001 in per side and the factor is 0.001/0.0014, or 0.7 (see Fig. 8-21). Excellent systems with tight process controls can produce one-to-one correspondence between artwork and conductors; that's because the slight flare which occurs in imaging compensates for undercut. A more conservative approach is to estimate 0.001 in of width loss per 0.0014 in of foil thickness; this factor is introduced at the design stage (see Chap. 4, "Design").

Etchants dissolve metal through an oxidation reaction at a rate which depends on solution activity and agitation. Some example equations follow.

Ammoniated cupric chloride:

$$\text{Etch: } Cu(NH_3)_4 + Cu \rightarrow 2Cu(NH_3)_2 \tag{8-1}$$

$$\text{Regeneration: } 4Cu(NH_3)_2 + 4NH_3 + 4NH_4 + O_2 \rightarrow 4Cu(NH_3)_4 + 2H_2O \tag{8-2}$$

Cupric chloride:

$$\text{Etch: } CuCl_2 + Cu \rightarrow 2CuCl \tag{8-3}$$

$$\text{Regeneration: } 2CuCl + 2Cl \rightarrow 2CuCl_2 \tag{8-4}$$

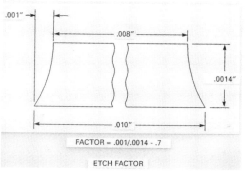

Figure 8-21 A conductor cross section showing how etch factor is calculated.

Ferric chloride:

$$\text{Etch: } 2FeCl_3 + Cu \rightarrow CuCl_2 + 2FeCl_2 \quad (8\text{-}5)$$

Spray impingement onto conductor sidewalls is inhibited by the overhanging resist resulting in slower lateral etch, thus less undercut. Further protection from undercut is provided by etchants such as cupric chloride which produce insoluble by-products as the first stage of their reaction [see Eq. (8-3), oxidation of copper from the FPW panel to cuprous chloride]. Cuprous chloride is a whitish residue which settles in the relatively quiet areas along the sidewalls, protecting them from further attack until additional chlorine is provided from the bath [Eq. (8-4)] to oxidize it to cupric chloride, which is soluble. The rate at which Eq. (8-4) runs is determined by the amount of free chlorine in the solution; for best etch factor rather than greatest etching speed, the bath is operated in a slightly "tired"—low chlorine—condition.

Something like this action is accomplished in ammoniacal etchants by operating in a copper-rich, ammonia-starved mode and by addition of banking agents to the replenisher solution.

Etcher controls

Uniform etch rate is important for quality work output. Production etchers are equipped with automatic measuring and feeding systems that are intended to keep the bath within a process window that yields constant etch rate and acceptable undercut. None of these systems is perfect; they all assume steady-state operation with consistent copper loading and replenisher chemistry, and variations in any of these factors will lead to out-of-tolerance product.

Ammoniated cupric chloride is maintained by measuring the specific gravity, which rises as copper is dissolved, and introducing replenisher (a mixture of water, ammonia, and compensating chemicals balanced for normal operation) to lower it to the set point. For steady-state operation, this method works. However, operation of the ventilation and spray systems without etching copper—for example, while awaiting work—unbalances the system because airflow through the sprays strips ammonia from the solution, lowering the pH without affecting the specific gravity and thus not activating the automatic replenishment system. When the operator detects this condition by the slowing etch rate, ammonia is manually sparged into the etcher sump until the pH rises to a desired set point.

Normal specific gravity settings range from 1.2 to 1.228 (roughly equivalent to 20 to 23 oz/gal of copper), and pH from 8 to 8.8. For fine line work, the solution is operated close to 22 oz/gal and at a pH close to 8. This condition will reduce etch rate but improve etch factor by almost 50%. Close attention is required, because it's risky to run with a pH which is too low; copper may crystallize out, necessitating a difficult cleanup process.

The status of cupric chloride baths is determined by oxidation-reduction potential and replenished by use of hydrochloric acid or chlorine gas additions to provide free chlorine, oxidized either by operating the sprays without workload or by use of hydrogen peroxide. Typical operating parameters are

Oxidation-reduction potential (ORP)	540 mV ± 40
Free acid	1 N ± 0.2 N
Temperature	120°F ± 2°
Specific gravity	1.2832

For fine line work, the ORP is allowed to fall to 500 mV and free acid to 0.8 N or less. *If hydrogen peroxide is used, it is critically important to maintain adequate free acid to avoid violent reactions.*

Electric etching

An unusual method of metal removal, electric etching is essentially a reversed plating process. The metal to be etched is imaged with an appropriate resist pattern, then exposed to plating current in a strong sodium nitrate electrolyte solution. The work is anodic; current flow is very large, with a solution rate of 0.1 in^3 of metal per minute at 1000 A as governed by Coulomb's law. Assuming 15 V across the cell, a typical 18- ×24-in FPW pattern with 50% of the metal removed, in 0.0014-in foil, would lose about 0.3 in^3 of metal in 3 minutes. Energy consumed is 0.75 kWh or around $0.10 at industrial rates corresponding to an attractive $0.07/oz of dissolved copper.

This process is simple and inexpensive in concept but considerably harder to operate in practice. Two major problems are holding the laminate in place close to the cathode against the solution agitation and avoiding the danger of cutting off or isolating areas of metal from the etching current before they're fully removed. The best procedure is to secure the laminate around a slowly rotating drum by light tension and remove the metal in a narrow line by exposing it to a small cathode area. This technique provides the necessary hold-down force and avoids the isolation problem.

Stripping

Resist coatings are consumables; they're applied and removed. The process of removing is called *stripping*. All currently used resists are designed to be stripped by water-based, or aqueous, chemistries in response to environmental demands. For most resists, the stripper solution is a more powerful form of the developer, usually operated at higher temperature; for example, 2% sodium hydroxide at 130°F.

Equipment for resist stripping could be identical to developer equipment: a conveyerized machine with spray chambers that apply the chemistry to both top and bottom surfaces, followed by rinse and dry stations (see Fig. 8-14).

Outline

The process of separating FPW from the panel and creating the desired border contours or outline is not as simple as might appear and occurs several times for each piece, when operations for creating coverlayers, stiffeners, bondplies, and the like are included. There are levels of automation and accuracy starting with hand-scissor cutting, proceeding to steel-rule dies either eyeball-aligned or pin-registered, and ending with high-speed, pin-registered multicavity outline dies.

Two sets of dimensions apply: the shape and size of the outline, and its registration to the etched pattern. As noted in Chap. 4, "Design," typical minimum edge distances between outline and etched pattern range from 0.02 in for steel-rule die cuts to 0.003 in for precision matched dies. Wider separations are strongly preferred to assure isolation of the outer conductor, particularly when a coverlayer is used, because the act of outline cutting tends to shock and fracture adhesive bonds (or covercoat integrity), leading to potential failure.

Figure 8-22 gives tolerances and life expectancies for three types of dies.

Tooling

Each FPW design requires many custom tools and an enormous amount of engineering and documentation. The up-front cost and time required is a major impediment to wider utilization of the product. Included in the tooling package are

1. Artwork—usually as a multi-up composite with coupons

Die type	Outline tolerance, in	Life, hits
Steel rule, eyeball-aligned	0.015	5000
Steel rule, pin-registered	0.01	5–8000
Matched multicavity	0.005	>50,000

Figure 8-22 Characteristics of dies.

2. Drill programs
3. Outline dies
4. Electrical test (see Figs. 8-23, 8-24, and 8-25)

Also frequently required are:

Figure 8-23 Detail of three-up FPW test fixture. Tooling pins are at lower left and upper right. (*Courtesy Circuitest Services Inc.*)

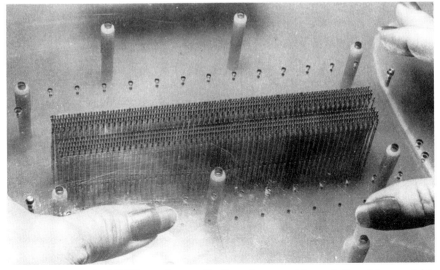

Figure 8-24 Detail of FPW test fixture. Tooling pins are at lower left and lower right. (*Courtesy Circuitest Services Inc.*)

162 Chapter Eight

Figure 8-25 Detail of FPW test fixture with FPW. Tooling pins are at lower left and upper right of the contact array. The FPW is rotated; tooling pin A is at lower left of the FPW, and tooling pin B is at upper right of the fixture. (*Courtesy Circuitest Services Inc.*)

5. Lamination fixtures
6. Potting tools, assembly, and forming fixtures
7. Nomenclature stamps and screens

The CAD station is fundamental in tool preparation. Common tooling hole and pad locations are called up from the database and introduced into all drawings and tooling for a design, assuring accurate alignment. In factories so equipped, CAD data is directly fed to drilling and routing machines, test stations, artwork plotters and NC machine tools for further labor reduction and error elimination.

Further details of tooling are set forth in Chaps. 3 ("Manufacture of Flexible Printed Wiring and Printed Wiring Boards") and 4 ("Design").

Artwork is a copy of the master pattern which is stored in CAD data form and replotted as needed. The life of a working copy is limited by buildup of scratches, handling damage, and distortion caused by vacuum draw-down over the FPW panels plus insertion and withdrawal of tooling pins. Each replotted copy requires inspection and touch-up before use. See Chap. 4 for discussion of artwork reproduction tolerances.

Drill and rout programs may be directly fed to the NC machines or supplied in floppy-disk format or in paper tape. In some cases a scaling factor can be applied to expand or shrink the hole pattern for best fit between drawing dimensions and etched pattern. Variable offsets from datum reference are common; they are determined by eyeball in single-layer FPW and by x-ray in multilayer circuitry.

Common tooling is not often found both because each design is unique and because tools are charged to the job, belong to the customer, and are kept exclusively for the customer's use. Generic lamination fixtures may be used—these consist of outer master plates which are drilled on a $\frac{3}{4}$- or 1-in grid, together with inner plates that expose only the desired hole locations for alignment pins. Master plates are house tooling, but the job buys the inner plates.

Outline configurations are determined as part of artwork design and design rules. Production of steel-rule dies is simplified by modem transmission of outline data direct to the vendor. Matched dies may be produced by direct NC equipment but more often require conventional drawings.

Electrical test fixtures are produced by both eyeball and CAD-based techniques; the choice is largely driven by complexity. Pad location is available from the database.

Assembly of FPW to hardware is frequently done by FPW manufacturers, in which case production of a variety of assembly tools and fixtures is required. Examples of such design-specific tools include:

Forming fixtures (for prebending)

Soldering fixtures (to position and secure hardware and FPW for mass or hand-solder attachment; they may include heat shields)

Potting molds (custom for FPW and connector, to customer designs)

Nomenclature screens or stamps

Prebending or forming is a difficult process which involves plenty of hand skill and subjective inspection. Where tight bends or creases are needed to fit the FPW into an assembly, simple jigs and fixtures are well worthwhile; these align to etched marks in the circuit pattern and function to limit the smallest formed radius. As most FPW material systems are at least somewhat thermosetting, forming can succeed only to a limited degree and will inevitably lead to springback (loss of shape) to varying extents. Large nonfunctional copper areas at the bend will help to preserve formed shape, as will adhesives or permanent clips or reinforcements. In thermoplastic systems such as polyesters, vinyls, and fusion bonds, better forming is possible; the circuitry is shaped and held by a fixture, exposed to temperature slightly above the dielectric softening point (by water bath or circulating oven), then quenched in cold water. Acrylic systems are noted for undesirable creep or flow if subjected to tight bends for long periods of time at slightly elevated temperatures. The best practice is to cold-form to a tighter shape than required, then assemble with sufficient loopage to avoid kinks or creases. Figures 11-3, 11-8, and 11-10 are examples of forming fixtures.

Alignment

Alignment or registration of tool to pattern is a fundamental process in FPW production. Registration is required for at least these steps:

Image

Throughhole drill/punch

Coverlayer/covercoat

Outline

Nomenclature

Multilayer lamination

Electrical test

Most tools or operations are aligned by pinning through tooling holes; see Figs. 8-26 to 8-28. Tooling holes are produced as a first manufacturing step by drilling or punching through unprocessed laminate or at a later point by eyeball or automatic alignment to etched fiduciaries. There are innumerable schemes for improving the alignment of tool to circuit, but all are affected by the fundamental instability of FPW materials. As noted elsewhere, dual criteria apply: termination pads must have concentrically located throughholes, coverlayers, or covercoats, and must also be on correct coordinates in order to align with assembly hardware. If throughholes are pierced in a finished part (i.e., after the last thermal exposure), they will be at correct coordinate position but concentricity of hole to etched feature will suffer. If throughholes are created early in the process, concentricity of hole to feature is improved but both will shift during subsequent processing and may be locationally out of tolerance in the finished part.

One method for reducing misalignment caused by material movement is *postetch punching,* accomplished by automatic equipment which determines the center of an etched fiduciary mark and punches a tooling hole at that location. See Figs. 8-28 and 8-29. Figure 8-28 shows a robotic optical punching machine which follows a program to the approximate location of a fiduciary, locates and punches the exact center, then follows the program to the next fiduciary and so forth. Figure 8-29 shows three optically punched tooling holes surrounding

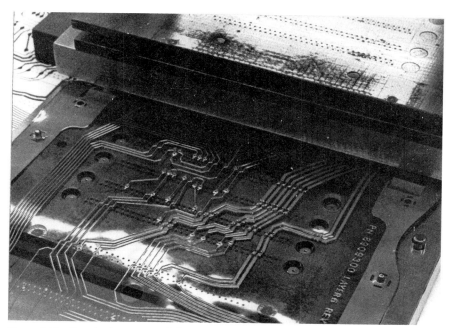

Figure 8-26 Punched pattern of clearance holes and termination holes. Two smaller pins align the FPW to the die. The large pin aligns the top half of the die (at top of picture, inverted) with the bottom half. (*Courtesy Teledyne Electronic Technologies.*)

a connector area. Even more sophisticated equipment aligns for best fit to all tooling holes in a cluster as a set rather than for each individual hole; this improves accuracy and reduces distortion and possible circuit damage when tooling pins are forced through slightly off-location tooling holes.

Cluster alignment (also called *multiple datums*) is an excellent technique for neutralizing the effect of FPW shrinkage, but customer drawings must reflect the same philosophy. The idea is to apply close tolerances only over an area where they are required, for example, within a pattern of connector holes; these holes are punched as a precise group by a tool registered to local or cluster alignment holes, and so forth for all required clusters. Distance between clusters is given a larger tolerance to reflect the fact that FPW is always designed to be slightly longer than the theoretical distance.

Process Sequences

At this point the reader should have an adequate grasp of the unique problems associated with manufacturing printed circuitry from flexi-

Figure 8-27 Close-up of pinning. Two pins protrude from the die base. The small-diameter pin engages the optically punched registration hole in the circuitry. The larger pin engages the top half of the die. (*Courtesy Teledyne Electronic Technologies.*)

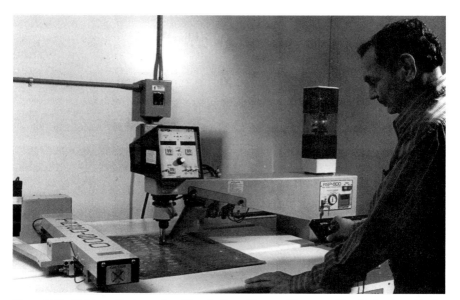

Figure 8-28 Ushio optical punch with robotic guidance. (*Courtesy Flex Technology Inc.*)

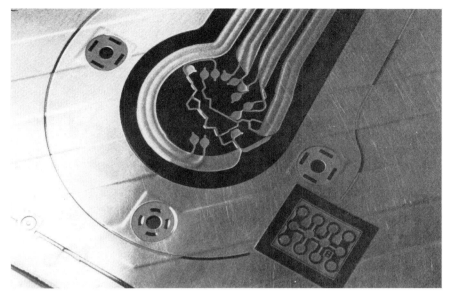

Figure 8-29 Close-up of cluster registration. Three optical fiduciary patterns have been automatically centered and punched. Shielded circuitry and ventilation pattern are visible in the embossed foil. The rectangular pattern with the nomenclature B is a test coupon. (*Courtesy Teledyne Electronic Technologies.*)

ble materials to appreciate and understand the reasoning behind the process sequences which are used. With this understanding, let's examine some of these processes and process sequences, always remembering that they're starting points which have many variations caused by equipment limitations, special product requirements, and so forth.

Single-sided FPW processes

Figure 8-30 shows the process flow for single-sided FPW. This product has one conductor layer supported on a base dielectric and may or may not have coverlayer or covercoat. Single-sided FPW is produced in the greatest production quantities of all FPW and is lowest in cost per conductor. The process sequence is relatively straightforward and is adaptable to the roll-to-roll production technique, which leads to even greater cost efficiency. Variations in dielectric and conductor parameters allow production of a wide range of product from undercarpet power distribution cables to high-density tape automated bonding (TAB).

168   Chapter Eight

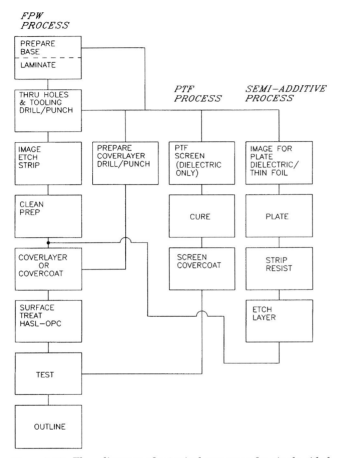

Figure 8-30 Flow diagram of a typical sequence for single-sided FPW with PTF and semiadditive alternatives.

Generic process sequence

1. Prepare base laminate (dielectric with conductor foil)
2. Drill/punch tooling holes
3. Image/etch conductor pattern
4. Apply coverlayer/covercoat
5. Drill/punch throughholes
6. Surface finish
7. Outline

8. Inspect/pack/ship

The basic process outlined above may be varied by changing the sequence of process steps. Some of these variations are

- Throughholes (step 5) may be created simultaneously with tooling holes (step 2). Combining these steps provides the advantage of a single drilling (or punching) step, but has the disadvantage of less precise throughhole location because etch and coverlayer processes will distort the FPW, upsetting hole-to-hole location.
- Steps 6 and 5 may be interchanged; i.e., surface finish may be applied before throughholes are made. For example, where solder is the protective coating, if it is applied after the holes are punched the diameter of the holes will be reduced. Reversing the sequence to solder-coat first, then punch provides better diameter control.
- Coverlayer/covercoat process may be deleted; some FPW applications, like PWB applications, are satisfied by uninsulated conductors.

Variations: upside down and reverse bare. By convention, FPW is drawn and conceptualized as though the exposed surface of the pads faces the viewer; the base dielectric is assumed to be the far side and the coverlayer the near side as viewed. This is in agreement with most commonly used manufacturing sequences and explains the ideation behind the term *reverse bare* for openings through the base layer. A fundamental goal in FPW manufacture is to create a circuit in which throughholes, pads, and coverlayer openings are concentric and at the desired coordinates. The difficulty in reaching this goal springs from the facts that (1) different process steps at different points in the sequence are used to create each of these elements, (2) elevated temperatures and stresses are involved, (3) the materials are dimensionally unstable, and (4) there's uncertainty in registering tools.

The upside-down process yields the best possible coverlayer-to-pad and pad-to-throughhole concentricity because all are generated and aligned at one process step. However, all will shift off location as a result of subsequent thermal processing, and the concentricity of clearance holes in the base layer to throughholes in the pad will suffer both for that reason and because of added alignment/tooling problems.

The upside-down sequence is

1. Prepare coverlayer with tooling holes.
2. Prepare foil with throughholes and tooling holes.
3. Laminate coverlayer and foil in alignment. Coat apertures with etch resist.
4. Image/etch (mirror image); strip.

5. Drill/punch base dielectric (clearance hole diameter should be slightly larger than that of termination holes to allow for misregistration)
6. Laminate full base dielectric in register to etched circuit
7. Surface finish, inspect, outline, etc.

The advantages of the upside-down process are better coverlayer aperture/pad/throughhole alignment. The disadvantages are that hole-to-hole dimensions will be degraded, aperture resist coating is required, and pads may be damaged from etchant leaks and additional process exposure. See Fig. 8-31.

A variation on the upside-down process, reverse baring is a method for creating apertures in both the base side of FPW and the coverlayer. It avoids the problems of postlamination baring, but at added cost compared to single-sided FPW. The decision to use reverse baring

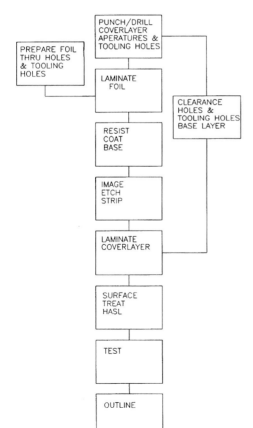

Figure 8-31 Upside-down process. The process manufactures FPW on a laminate of coverlayer and foil, upside down from the usual process.

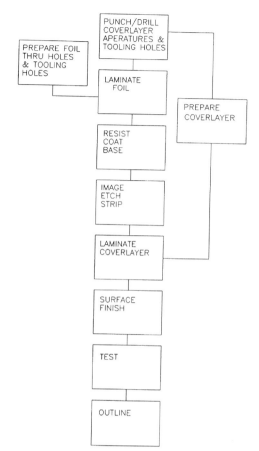

Figure 8-32 Reverse-baring process. Custom laminate has apertures in the base dielectric film.

depends on the number of openings per circuit—if there are just a few, laser ablation or mechanical skiving could be less expensive.

The reverse-baring process sequence is

1. Prepare base dielectric with backside apertures and tooling holes.
2. Prepare and laminate foil to it.
3. From tooling holes, apply etch resist to the foil side in register with backside apertures. Protect backside apertures with additional resist coat.
4. Etch/strip resist.
5. Prepare and laminate coverlayer.
6. Drill/punch throughholes.

The work is then completed as generic FPW. See Fig. 8-32 for the process flow.

Alternative imaging processes

The subtractive process is the most universal technique for creating FPW conductor patterns. The resulting circuitry has good electrical performance, is available in a range of thicknesses (thus resistivities and current-carrying capacities), is easily terminated by all techniques, allows field repair or rework without special equipment, and can be built at reasonable cost. Significant quantities of FPW are also produced by alternative imaging processes such as semiadditive or polymer thick film (PTF) processes.

Semiadditive process. This technique utilizes a seed coated dielectric film—a film which is very lightly metallized to provide a well-adhered, conductive substrate for subsequent electrolytic buildup of the conductor pattern. Conductors and pads are defined by openings in a layer of photoresist which is stripped off after the desired thickness has been reached. The seed layer is then removed by a brief exposure of the completed circuit to mild etching—the amount of metal removed has little effect on the conductor pattern.

The semiadditive process is rapidly gaining market share for designs which require extremely fine features—0.003-in lines and spaces or less. This technique uses "adhesiveless" materials which are readily available with the required seed layers and is particularly cost-effective for double-sided PTH circuits when throughholes are punched or drilled through the dielectric film prior to adhesiveless metallization. Seed layer deposition then occurs on both surfaces plus the holes so that the resulting custom material need only be imaged and etched (if plated to full thickness by the laminate supplier) or plated up (if purchased at the seed-layer stage) to create a double-sided, PTH-interconnected circuit. Semiadditive laminates have seed layers nominally 200 μin thick.

The semiadditive process sequence is

1. Prepare base laminate (dielectric with seed layer).
2. Drill/punch tooling holes and throughholes.
3. Image for plate. Plate to desired thickness.
4. Strip image. Differential-etch to remove seed layer.
5. Covercoat, etc.

The advantages of the semiadditive process are that it produces very fine features (less than 0.003-in) with better precision than etched runs because the deposit builds up in the resist image, which controls the sidewalls, and it has best dimensional stability because the metal/dielectric composite is formed without lamination under heat and pressure. Its disadvantage is that it requires careful control of plating conditions to yield conductor patterns with uniform plating thickness and good flexural properties. Figure 8-30 shows semiadditive and PTF flow.

Polymer thick film or conductive ink printing. For further discussion of PTF materials and technology see Chap. 6, "Conductive Materials." The process sequence for PTF is

1. Prepare base laminate (dielectric only).
2. Print conductive pattern and fiduciaries.
3. Cure conductor pattern. Punch tooling holes from fiduciaries.
4. Print dielectric (covercoat) layers.
5. Outline, etc.

PTF's advantages are that it reduces toxic waste generation and lowers cost; it also eliminates foil and base lamination processes for simplified processing and better stability. PTF's disadvantages are high resistivity, low current-carrying capacity; it also requires special assembly techniques (conductive adhesive or reduced pressure) because conductors are not solderable.

Die stamping. This is a method used for low-cost, high-volume production of somewhat coarse-featured FPW. There are several variations, all based on the concept of forming or coining the circuit pattern by mechanical deformation of copper with a hardened tool. The most common process consists of cutting areas out of an adhesive-coated foil with knife-edged die and transferring them to a dielectric sheet. The die is heated and activates the adhesive enough to cause the desired part of the foil—inside the cutting area—to adhere to the dielectric as the scrap is pulled away.

Many automobile dashboard circuits have been built by this process, which is adaptable to roll-to-roll methods and capable of producing conductors 0.012 in wide or more. It must be realized that all mechanical processing of copper, particularly annealed foil, causes work-hardening; even the slight flexing and elongation which unavoidably accompanies pumice cleaning causes loss of ductility. Mechanically formed circuits should not be used where flexing is required; they are not functionally equivalent to circuits etched from rolled-annealed (RA) foil.

Sculpting. The Sculptured* flexible circuit process is step etching applied to flexible circuitry; process flow is essentially identical with

---

*Registered trademark of Advanced Circuit Technology, Inc.

methods used for producing lead frames. The principal benefit of this process is that it allows variations in conductor thickness, for example, thick at termination points where strength is needed, and thinner in sections where flexibility is required.

The most common use for this specialty process is to generate cantilevered fingers extending beyond the edges of the coverlayered circuit as built-in terminating hardware. Terminal pads can also be manufactured by the process. Pads are created in thick metal and therefore are particularly rugged. They protrude through the coverlayer to lock it into alignment with the etched circuit and have outstanding solderability because they're above the dielectric, thus easily cleaned.

Throughholes are produced by etching and are very well centered in the terminals because they are artwork-controlled. There's no material instability to contend with; the two etch images are applied to opposite sides of a solid foil sheet. Concentricity of hole to pad is thus as good as artwork alignment, with roughly a 0.002-in tolerance. Terminal pads are created early in the process and extend through the base dielectric, something like the upside-down process. See Fig. 8-33.

Greater shrinkage after coverlayer application is common in this construction because of increased distortion of dielectric films around

Figure 8-33 A circuit with chemically milled termination hardware integral with conductor runs. The step etch technique was used; a step is visible beneath the covercoat. The nominal termination thickness is 0.01 in; conductor thickness is 0.004 in with a tolerance of 0.001 in.

the thicker foil terminals. Because the foil is much thicker than standard, the ability to create fine-pitch features is limited. Cost is greater than standard FPW because more process steps are required, and much of the initial foil is removed and discarded by the etching steps.

Double-sided FPW processes

Direct access. The process sequences used for single-sided FPW with direct termination to exposed pads can be expanded to produce double-sided circuitry, also terminated directly to exposed pads. For example:

1. Prepare base laminate (double-clad, i.e., foils on both surfaces).
2. Drill/punch tooling holes.
3. Image/etch sides 1 and 2 (in register).
4. Coverlayer/covercoat.
5. Surface finish.
6. Drill/punch throughholes.
7. Inspect/test, etc.

This produces a circuit which is assembled from two sides, a cumbersome technique.

An alternative process which avoids two-sided assembly:

4. Machine/ablate/etch side 2 apertures through the base dielectric layer.
5. Full coverlayer side 2; coverlayer side 1 with all apertures for sides 1 and 2.
6. Drill/punch throughholes, etc.

Plated throughhole processes. PTH is the preferred technique for multilayer FPW. There are many process sequences; see Fig. 8-34 for generic process flow. Added steps required for PTH are hole drilling, plasma treatment, and electroless deposition. These processes account for a great deal of the troubleshooting and analytical workload in FPW production; FPW engineers spend many hours studying cross sections of PTH holes to verify that hole quality, determined by absence of nailheading and adhesive smear onto inner lands, is acceptable. Achieving this goal is particularly difficult in constructions with more than one layer, where interfacial bonding is necessary. The mixture of materials in an FPW cross section is almost completely wrong for good drilling, and the PTH process, given the variety of materials, is also challenging.

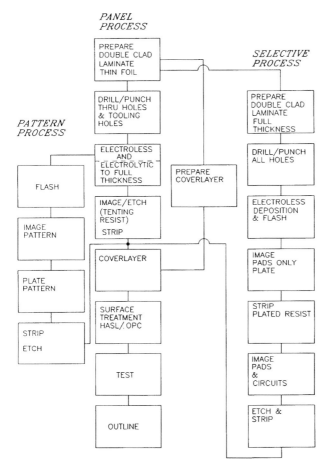

Figure 8-34 Flow diagram of basic steps in the PTH process with panel, pattern, and selective variations.

PTH drilling. Every material used in FPW is hard to drill and the combination is even worse. Copper is a difficult metal to machine because it's very soft and yielding. The plasticized adhesives in FPW are similarly mushy; they provide poor support and in addition melt at drill-generated temperatures, which means the hole walls are smeared with displaced adhesive. FPW dielectrics are also uncooperative in that they either melt readily (polyesters) or are extremely tough (polyimides).

In general, correct conditions for drilling FPW for PTH are similar to those used for coverlayers: high infeed to keep the drill constantly moving into fresh, cool material; good entry and exit materials; 300 to 400 SFM tip speed; and good tools. For example:

Drill diameter: 0.032 in

Chip load: 2.5; infeed 112 in/min

Speed: 45000 r/min

Stack: up to 6

Entry: 0.012-in in hard aluminum foil

Backup: 0.093-in aluminum-faced particle board. If needed, insert additional sheets of entry material between third and fourth panels.

Minimize the force applied to minimize distortion with few layers or thin constructions.

Hand pumicing, a process used to clean and deburr after drilling, is performed at stations like that shown in Fig. 8-35.

Plasma. Plasma treatment is about the only option for removal of smeared adhesive from the inner land surfaces. Chemical treatment won't cleanly remove adhesives and presents disposal and control problems. The plasma process is extraordinarily complex, involving high vacuum, radio-frequency (RF) power, and flow of mixed, reactive gases. Panels of FPW are baked for 1 to 2 hours immediately prior to treatment to thoroughly remove moisture, then racked into the machine as shown in Fig. 8-36. A cycle requires somewhat more than an hour and consists of

- Heatup in inert gas with RF energy to raise the panel temperature to a suitably active point.

Figure 8-35 Hand pumicing station has air-driven rotary brush, rinse hose, and black elastomeric support that cushions and grips the FPW sheet while it is cleaned. (*Courtesy Teledyne Electronic Technologies.*)

Figure 8-36 A plasma machine. An operator is inserting a panel into the left side of the two-chamber machine; the left cover has been slid back. Controls are at left; a large vacuum pump and a radio-frequency power supply are at rear. (*Courtesy Teledyne Electronic Technologies.*)

- Treatment in active gas (carbon tetrafluoride and oxygen) activated by several kilowatts of RF power, continued until the panel temperature reaches a maximum limit (90 to 100°C), followed by cool-down in oxygen without RF energy.

Plasma treatment oxidizes organics to an ash which is removed by subsequent wet process steps but doesn't affect metal such as the outside foil surfaces or exposed inner lands. It is a touchy procedure which is unbalanced by moisture, erratic field and temperature buildup (edges and corners are always attacked more strongly), and depletion of active species. But, if done correctly, plasma results in very clean inner land surfaces and etched-back organics, i.e., the pad surfaces protrude slightly into the hole.

Electroless deposition. This magical process applies a thin but conductive layer of copper to all the outer surface of the panel including hole walls. In simple terms, two interacting baths are used to deposit a noble metal—usually palladium—which is then replaced by copper.

The process requires as many as eight different chemical paths with multiple intervening rinses. An electroless deposition line is a delicate thing to keep operating; constant analysis and adjustment or replenishment is needed, as is constant loading. An enormous amount of rinse water is consumed in the process—perhaps a dozen rinsing operations, which translates into expensive supply and wastewater treatment. See Fig. 8-38.

A representative electroless deposition process sequence consists of the following steps:

1. Glass etch (if glass-reinforced PWB materials are included); ammonium bifluoride 30% at 120°F; 3 to 5 minutes.
2. Two water rinses.
3. 10% sulfuric acid neutralization.
4. Water rinse.
5. Cleaner/conditioner (prepares adhesive for electroless deposition).
6. Water rinse.
7. 2% sulfuric acid rinse.
8. Water rinse.
9. Conditioner (different from step 5).

Figure 8-37 Measuring the plating thickness on a panel of circuitry. Note the rack. (*Courtesy Teledyne Electronic Technologies.*)

10. Water rinse.
11. 2% sulfuric acid rinse.
12. Water rinse.
13. Microetch; 50 μin copper removal.
14. Water rinse.
15. 10% sulfuric acid rinse.
16. Water rinse.
17. Predip (catalyst but without palladium).
18. Catalyst.
19. Two water rinses.
20. Accelerator.
21. Water rinse.
22. Electroless deposition/flash electroplate.

Figure 8-34 shows a typical manual plating line.

A major concern in electroless deposition is that there's always a weak layer of electroless copper separating the inner pad lands from the barrel; the interface can fail in thermal stress, causing product rejection. This weakness is eliminated by *direct metallization* processes now available which also simplify and reduce the cost of the PTH

Figure 8-38 Electroless deposition line. (*Courtesy Teledyne Electronic Technologies.*)

process. These newer hole metallizing technologies involve fewer process steps, much less rinsing, and direct attachment of electrodeposited barrel copper to the inner lands. As industry confidence builds with continued use, direct metallization is likely to replace electroless deposition for commercial production.

Flash plating. Some electroless deposition baths produce a highly stressed metal layer which will self-destruct by peeling away from the panel if allowed to accumulate to more than a few microinches in thickness. Therefore, after the holes have been metallized by either electroless or direct metallization, the fragile deposit is reinforced by flash plating with a light electrodeposited layer to preserve the interconnection and to build up enough copper for storage and cleaning prior to either photoresist application (for the pattern plate process) or reactivation (for panel plating). Some metallization methods will build sufficient thickness for subsequent processing without flash plating, provided imaging and electrodeposition take place quickly.

A cross section to verify hole and electroless deposition quality is normally taken from a pilot panel before a production lot is committed.

Process sequences for double-sided FPW. The basic panel process sequence is

1. Prepare base laminate (dielectric plus foil).
2. Drill/punch tooling and PTH holes.
3. Electroless-deposit and plate to desired thickness.
4. Image/etch (resist must tent to protect PTH barrels).
5. Prepare coverlayers; align and laminate.
6. Surface-finish; hot-air solder level; apply organic protective coating.

An alternative method produces the conductor pattern by pattern plating on a semiadditive (seed layer) laminate:

3. Electroless-deposit and flash-plate.
4. Image to expose conductors and pads.
5. Plate to desired thickness.
6. Strip image; differential (brief) etch to strip seed layer.
7. Surface finish, etc.

Another alternative uses metal etch resist:

5. Plate to desired thickness; finish with tin-lead solder.
6. Strip image; etch, using tin-lead as resist (ammoniacal etchant).

7. Strip tin-lead.
8. Surface finish, etc.

Both alternative processes produce FPW which has vendor-deposited copper on the conductor surfaces. The clad or initial foil can be a thin seed layer since plating the PTH barrels to required thickness will also plate the same thickness on the conductor runs.

A description of the adhesiveless technique for producing double-sided PTH in high volume is given in the section "Semiadditive Processes" above.

Where specifications require RA foil conductors, neither the panel nor pattern process can be used. In those cases, the alternative is the selective plating process:

1. Prepare base laminate (dielectric plus RA conductor foil).
2. Drill/punch tooling holes and PTH holes.
3. Electroless-deposit and flash-plate.
4. Image to expose pads only; plate to desired barrel thickness.
5. Strip plating resist image; reimage pads plus conductors (tenting resist protects barrels), and etch.
6. Prepare coverlayers. Apertures are sized to accept plated pads but cover the clad foil pad.
7. Laminate
8. Surface-finish, outline, etc.

The advantage of selective plating is that it allows thin, high flexural endurance RA conductors with thicker PTH pads and barrels. Its

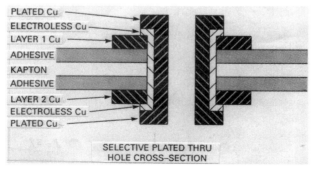

/**Figure 8-39** Cross section of 1-oz RA clad foil with PTH interconnection by the selective PTH process. Doubled copper thickness at the termination can complicate coverlayering.

disadvantages are that two image steps are required and there is danger of etch undercut at the PTH pad/conductor run interface where the abrupt change in surface may affect the dry film resist seal. Figure 8-39 shows the details of a PTH hole without coverlayer.

Summary

A wide range of process conditions and equipment are employed in building FPW. The flexible substrate allows roll-to-roll processing in those facilities which have the necessary high-cost equipment. Most FPW is built by the panel process, which reduces start-up tooling cost and allows a wider range of process conditions.

Equipment used in FPW production is essentially identical to PWB equipment. The thin, weak materials cause considerable added labor in special handling and use of leaders with conveyerized equipment.

The coverlayer process is unique to FPW; it involves preparation and lamination of a second dielectric film onto the completed conductor patterns. A covercoat—similar to a solder mask in PWBs—is also employed but requires flexible polymers with reduced cure temperatures.

All FPW processing is complicated by the dimensional instability of the materials. Alignment and registration of tooling and added layers in multilayer construction require careful design and added engineering; postetch punching provides improved alignment by reducing the effect of postetch shrinkage.

A wide range of production sequences are used to produce even single-layer FPW. More sequences are needed for double-sided and multilayer constructions.

The PTH process provides increased circuit density and simplifies design approach to connectors.

Chapter

9

Rigid-Flex Manufacture

Introduction

Rigid-flex is a confusing, oxymoronic contraction. Spelled out, it makes more sense: *combined rigid and flexible circuit.* A rigid-flex (RF) circuit could be described as "a printed wiring device which has areas of rigid and flexible dielectric with coextensive conductors throughout." A typical RF circuit consists of multiple layers of circuitry, bonded together in rigid areas and interconnected by PTH.

FPW with stiffener (see Fig. 9-1) is the simplest form of RF; Fig. 9-2

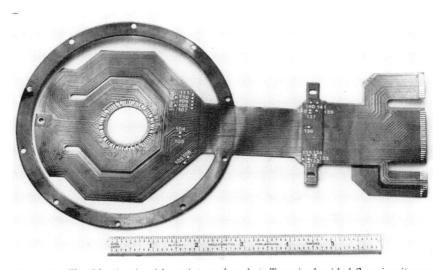

Figure 9-1 Flexible circuit with an integral gasket. Two single-sided flex circuits are bonded together only in the ring stiffener and strain relief areas.

Figure 9-2 Military rigid-flex motherboard for cannon-launched "smart shell." The motherboard resists extreme shock. Eleven polyimide-glass hardboard areas are interconnected by multiple thicknesses of flexible circuitry. The product is manufactured in high volume.

shows a very complex one. Both have the RF characteristic of rigid areas and flexible areas with interconnecting, coextensive conductor layers.

Rigid areas serve as strain reliefs, stiffeners, and component supporters; flex areas interconnect between rigid areas, absorb slight locating errors and provide limberness for assembly. RF circuitry satisfies virtually every requirement of electronic packaging in one structure. It takes longer to design, is harder to build, and more costly to procure than other PW forms, but RF is the most dense, versatile, and reliable interconnection available.

This chapter is dedicated to a clarification of the added manufacturing steps required for RF. At this point in the text, flexible printed wiring (FPW) should be a familiar product. The reader knows how FPW is built, the materials and techniques that are used, and the major manufacturing problems that result from the inherent nature of flexible laminates, coverlayers, and specified adhesives. That's the background for the following discussion.

Definitions

New terms and concepts come into use in RF, and a moment spent on introducing them will help clarify the intricate manufacturing procedures.

A typical RF board or circuit has two rigid *cap* boards—one on each outer surface—with one or more layers of FPW sandwiched between. Generally, but not always, outer surfaces of cap boards are bare except for pads which terminate the layer-interconnecting PTH. In some instances, *surface conductors* are found on the cap boards where they're exposed to short-circuiting and handling damage. Designs which include surface conductors may thereby avoid a circuit layer, but benefits of relocating surface runs somewhere inside the multilayer design almost always outweigh the slight additional cost. Cap boards and FPW *details*—the coverlayered layers—are solidly bonded together throughout rigid areas, which contain the PTH terminations. FPW layers may or may not be bonded to each other in flexible areas, and are sometimes single-clad and sometimes double-clad, with choice depending on the relative importance of flexibility (single-clad is most flexible) or manufacturing cost (double-clad is more cost-effective).

Cap boards are either preassembled as multilayer constructions, fabricated as standard double-clad PWB (foil-clad on the outer surface, etched pattern on the inner), or constructed during multilayer lamination by *foil* or *cap* lamination. Foil lamination consists of building up outer surfaces of the panel of RF circuitry with layers of prepreg topped by a sheet of foil. Cap lamination is a similar process except that the foil layer is replaced by a single-sided sheet of PWB laminate. Either process can be used in place of the conventional double-clad cap board process to accommodate odd-number layer counts or for other process-related reasons such as slot sealing.

Before the RF is laid up for lamination, cap boards, FPW details, *bag layers,* and *prepreg* or bonding adhesive are tooling-hole punched, *windowed,* slotted and partially outlined to create unbonded areas or edge contours which would be difficult or impossible to machine in the finished RF circuit. Windowing is performed with steel rule dies which register to adhesive or prepreg layers by means of tooling holes and pins to precisely cut out certain areas. The same process is used to create *fillers,* cut by the same die from sheets of *release* material—which are the same thickness as the adhesive—such as FEP Teflon,* Tedlar,† or TFE-glass cloth. Fillers are inserted or inlaid into the window areas during layup. Functions of fillers are

- Restore stack-up thickness for uniform lamination pressure
- Prevent adhesion between FPW layers
- Block adhesive (or prepreg) flow
- Minimize distortion

Cap boards are *slot-routed* along the edge where FPW emerges from the rigid area. If cap boards aren't preslotted (or scored on the

---

*Registered trademark of DuPont.
†Registered trademark of DuPont.

inner surface for breaking off) before lamination, cutting this edge in the finished product will require very delicate Z-axis control to avoid damaging the FPW.

Edges of FPW layers which won't be bonded together in the finished part, and therefore would be difficult to rout, are partially preoutlined by steel rule dies. There's no kerf, no scrap, nothing is removed. The FPW layer goes into the makeup as a complete sheet, borders, scrap area, and all, to aid in alignment and thickness control.

Cap board slots must be sealed, which is the function of the bag layer, a membrane of polyimide film or bonding prepreg inserted in the stack-up between cap board and FPW, and extending across the slot. Film bags are sealed to the slot perimeters by over- and underlying adhesive layers; prepreg bags are self-sealing. Another technique uses foil lamination with windowed prepreg.

If the design requires PTH areas that have fewer layers and are thinner, a difficult process called *sequential lamination* is used. In this technique, those layers that terminate in the thinner areas are laminated and PTH-processed as subassemblies, then introduced into a final RF layup. At this point the already completed PTH areas are sealed inside the thicker RF panel with additional layers of FPW and cap boards, building up to finished thickness for a second PTH process.

Unbonded sections between rigid areas—the *pouch areas*—if large enough may expand and cause delamination at plasma treatment, depending on how well-sealed and free of volatiles they are. When pouch areas exceed 4 to 5 in^2, the expansive force generated in the hot vacuum plasma process may delaminate bordering edges of the cap board. *Vent holes* are drilled into the pouch to relieve this pressure and must be tightly sealed before the PTH process.

Stringent special quality controls apply to RF circuitry. Perhaps the most challenging inspection procedure is *thermal stress,* a visual and cross-sectional analysis performed on a representative PTH *coupon* cut from the panel. The coupon is baked for a minimum of 6 hours at 125°C, then cooled, fluxed, and floated on molten solder at 287.7°C (550°F) for 10 seconds. It's examined for surface imperfections such as weave texture, weave exposure, scratches, haloing, pits, and dents, then cross-sectioned and analyzed for plating integrity and a wide range of characteristics in the *A* and *B* zones. The A zone is the area immediately around a PTH, including contiguous pads and conductors; the B zone starts 0.003 in away from the nearest A zone conductor and extends to the next PTH location. A common reject cause in RF products is *laminate voids* in the B zone—these are voids or bubbles in the dielectric structure; a product is rejectable if laminate voids are larger than 0.003 in or if they violate conductor-to-conductor spacing rules. Figures 9-3 to 9-8 are cross-sectional photomicrographs

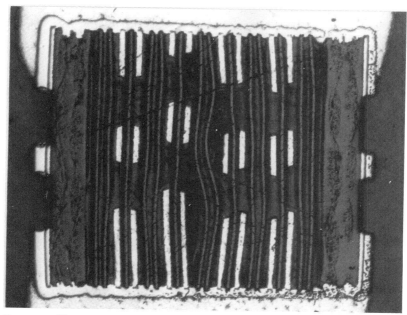

Figure 9-3 View of A and B zones of an acrylic-bonded MIL-P-50884 coupon after thermal stress. Three laminate voids are visible. The circuit contains four double-clad FPW layers and has film bags at each cap board (14 layers of 0.001-in polyimide film are visible). The large amount of adhesive and distorted polyimide films is evident.

of thermal-stress coupons. Figure 9-9 is a reproduction from MIL-P-50884 which further explains these inspections.

In applications which require very tight bends in the FPW, *progression* is used to reduce stress in the installed condition (but at the expense of greatly complicated manufacturing processes and high stress during manufacture and solder assembly). Progression is a design technique wherein FPW layers, proceeding from inside to outside of the bend, are made progressively longer to compensate for increasing path length. Progression is discussed in Chap. 4, "Design." Another term for progression is *bookbinder*, used because the arching FPW layers look like the spine of a book laid facedown with open pages. See Fig. 4-3.

Figure 9-10 shows a 14-layer military RF circuit flexed as it will be in the installed condition; note the tight bends that are required and the buckling in the six FPW double-clad circuit layers. Figures 9-11 and 9-12 show a bookbinder RF circuit in the flat (stressed) condition and formed 90° for installation; note how the FPW layers are now less stressed.

Figure 9-4 Detail of a thermally stressed coupon from an eight-layer RF circuit. The acrylic adhesive in the B zone has two laminate voids. The large amount of adhesive and distorted polyimide film is evident.

Process

It's been estimated that RF manufacturing has three times the number of process steps, thus opportunities for rejection, as multilayer PWB. RF construction also includes flexible laminates and adhesives which importantly misbehave at elevated temperatures, compared with PWB resin-glass systems.

RF manufacture demands extraordinary care at every step. Just as in the case of FPW, RF production carries the constant threat of undetected material defects or process deviation which rears up at a later process or inspection step to cause rejection. RF is a very high-value-added product: late-cycle rejection inflicts enormous cost in lost labor and time.

The most likely rejection point is final inspection; particularly the dreaded thermal stress analysis, which is the first time that the integrity of the FPW layers, cap boards, and dielectric system is tested. This is where a tiny 0.003-in void in the B zone causes rejection of product at its highest value, on the eve of delivery, with no hope of repair or rework. Days, if not weeks, of labor plus layers of FPW and cap boards which must be recreated from scratch are lost in a

Figure 9-5 Thermally stressed high-performance MIL-P-50884 coupon in prepreg construction (Zontar process). Films are flat; the cross section is unaffected by thermal stress.

moment. To minimize this problem, production of RF proceeds with higher-than-normal supervision and engineering oversight.

Overview

If necessary, now's the time for a quick review of the PTH process in Chap. 8. That section gives the steps that are involved in producing double-sided FPW; the same procedures and equipment are used in RF.

We'll assume that FPW and cap board layers are correctly designed to provide the desired interconnections and dimensions; that correct tooling, drill programs, artwork and material callouts have been provided; that the factory functions with good efficiency and tight process control. Up to the stages of premachining, layup, and beyond, RF production is identical to PWB and FPW production. In fact, most of the raw material for RF—etched layers, or details—is the output of PWB and FPW manufacture. From that point on, the special RF manufacturing occurs; in similar fashion, consider this chapter an addendum to Chap. 8, "Manufacturing Processes." From here on we deal with specifics of RF production.

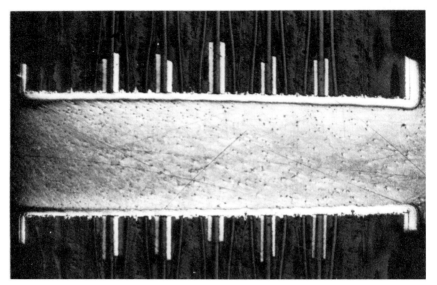

Figure 9-6 Thermally stressed coupon from a 12-layer RF circuit. The product consists of plated-up adhesiveless FPW layers with prepreg bonded coverlayers and prepreg bondplies (Zontar process). Etchback is inhibited. The cross section is unaffected by thermal stress.

Figure 9-7 Close-up of adhesiveless layer-barrel interface. The horizontal, multilayered grain structure in high-endurance ED foil is visible. The near-vertical pad edges show excellent etching properties. Coverlayers are prepreg bonded. Etchback is inhibited by absence of adhesive layer beneath the foils. The electroless-deposited interface is heavily stained.

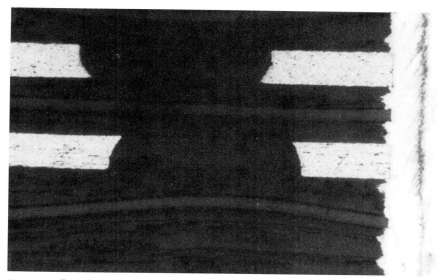

Figure 9-8 Detail of rolled-foil conductors in Zontar RF cross section. The grain structure in the rolled foil shows horizontal orientation. Etchback is improved because prepreg surrounds the pads.

Sealing

A panel of RF circuitry contains layers of etched details and adhesive aligned on a network of tooling pins. Unlike other multilayer PTH circuitry—PWB or FPW—layers inside a panel of RF circuitry are discontinuous; i.e., they have precut edges and contours and cutout areas. Unless the cutout areas are filled, the layers won't exist throughout the panel to help maintain uniform thickness, good sealing, and so forth.

Borders are always completely present and sealed to each other and to the cap boards by adhesive films or prepreg layers; internally, things are different. In order to allow unbonded areas between FPW, and to create contours around rigid edges from which the FPW will extend in the finished circuit, it's necessary to premachine, slit, slot, and window the FPW and adhesive layers. So that adhesive is removed from areas that aren't to be bonded and replaced by fillers for uniform thickness, contours that can't be cut in the laminated rigid-flex panel are already cut in the details, and so forth. In spite of all these material jugglings and machinings, the laminated panel must present totally sealed, fissure- and void-free cap board surfaces and border edges for plasma and wet processing. This is what drives the complicated windowing, filling, bagging processes—the layers have to be violated to create the edges of the final part, and then temporarily made sound again to withstand remaining process steps.

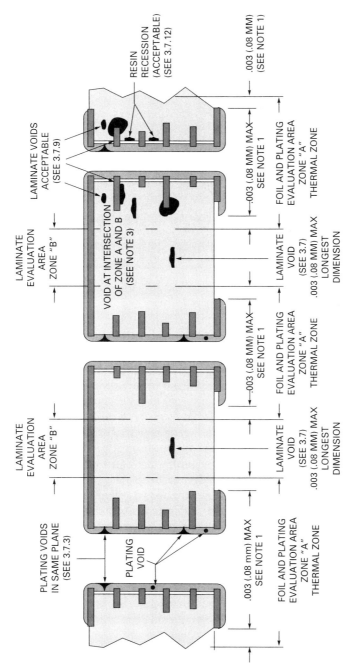

Figure 9-9 Extract from an early MIL-P-50884 specification illustrating A and B zones with limits for voids and other defects.

Rigid-Flex Manufacture   195

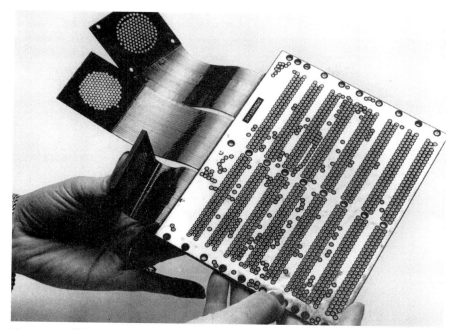

Figure 9-10 Military 14-layer rigid-flex board for an avionics applications. Flexure of the head patterns provides a 90° interconnection path and tolerance absorption. The main board surface has solder-coated ground planes; the mounting holes have strain relief.

Figure 9-11 Flat bookbinder RF circuit.

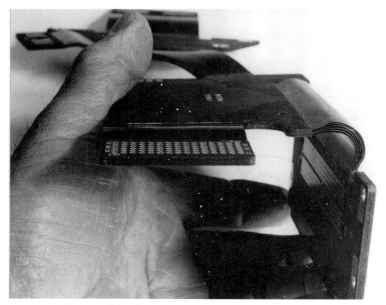

Figure 9-12 Bent bookbinder RF circuit.

Figure 9-13 is a schematic layout of the layers of the RF circuit shown in Fig. 9-10. The layout shows the slotted cap boards, windowed and filled adhesive layer, and preoutlined FPW layers.

Bags. Bag sealing is a patented process. The technique is used throughout the industry, but the reader should be aware that a patent has been granted for both prepreg and polyimide film bag constructions.

Picture a cross section of FPW and cap board layers, interleafed by adhesive layers. In flexible areas the adhesive has been cut out and replaced by fillers so that the total material stack-up is uniform in thickness. Bonding can't occur where it's not wanted—not between FPW layers, nor between FPW and cap board in the flex or pouch area, because fillers are cut from nonbonding release films. However, even after the rigid areas have been sealed together by the lamination process acting on the prepreg or adhesive layers, the interior of the pouch is full of fissures and blind pockets surrounding the FPW and fillers. The cap board slot overhangs this unsealed pouch area and thus is a leakage path through which PTH process chemistry can enter the interior of the panel with disastrous consequences for plating bath composition and circuit contamination. The slot must be sealed, but not in a way which is hard to unseal to re-create the desired contour in the finished part.

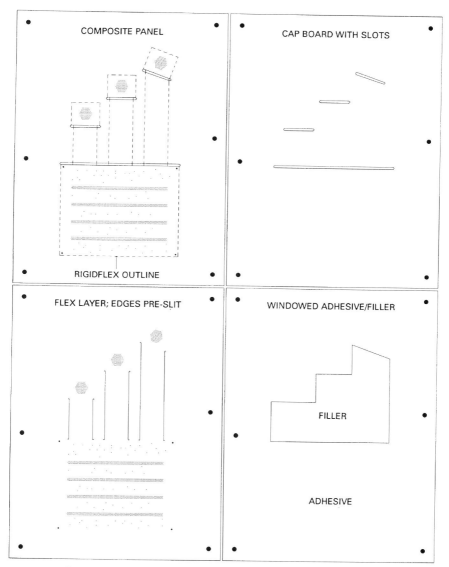

Figure 9-13 Schematic illustration of the layers of a rigid-flex board, showing the relationship among slots, rigid and flex areas, fillers and adhesive layers, and precut FPW edges. Note the common tooling pattern in all layers.

Bag layers provide this function. A bag is a sacrificial layer lying between the FPW layers and the cap boards. Bags are bonded to each cap board, covering its slots, by a full sheet of adhesive, but only bond to the adjacent FPW layer in rigid areas—the adhesive layer between bag and FPW is windowed and filled in the flexible area.

An alternative method uses the foil layer in foil lamination as a bag or slot seal. In this technique the slot is sealed on its outer surface by a sheet of foil bonded onto the cap boards by prepreg layers that are windowed out at the slot. Only the foil covers the slot, and it's chemically removed during final etch. A second, temporary seal may be required to protect the RF through solder reflow and final cleaning.

Vents

Pouch areas inevitably contain volatiles, even with the vacuum-assisted lamination which is mandatory with RF panels. When an RF panel is plasma-treated, the combination of elevated temperature plus vacuum creates the chance for delamination along the edges of the pouch area as the trapped volatiles expand, forcing upward against the cap boards. If the pouch is small and the panel is well-bonded, no damage will be done. But larger pouches—above 5 in^2—should be vented. To do this, a hole which breaks into the pouch is added to the drill program. The hole is left open through plasma treatment to equalize the internal and external pressures, then sealed with platers' tape, solder, or other means prior to wet processing.

Bookbinder

Bookbinder RF is devilishly difficult to process at many steps, of which sealing for the PTH process could be the worst. Where progression is minimal, it's possible to compress the added FPW length under the cap boards, thus avoiding added process problems. Typically, bookbinder cap boards are routed out to allow the longer, outside-the-bend FPW layers to bulge outward (through the cap board), relieving compression. The exposed FPW surfaces and complicated internal contours where they emerge from the rigid area must be sealed to prevent electroless deposition, carry-through PTH bath contamination, and board rejection from insulation contamination. Methods for sealing progression RF panels depend on the number of panels to be processed and range from skill-demanding hand taping with plater's tape to bolted-on, gasketed temporary enclosures to complicated, preformed plastic film bags which extend out through the cap board relief to encapsulate the bookbinder bulge.

Other bookbinder process problems:

- Each FPW layer grows progressively longer toward the outside of the curve. For example, layer 1 may measure 2 in from rigid area A to B; layer 2 may be 2.020 in between A and B, and so forth. The distance between area A and area B tooling holes is also increased, in each layer, by the same amount; at layup, when the layers are pinned in register, each FPW layer bulges upwardly according to its extra length. It's necessary to slit the FPW edges and cut out the scrap areas, including the borders, to allow free redistribution of the extra length and minimize the buckling of the layer. Careful die design is needed to assure that the cut edges of the borders butt back together to restore the edge seal.

- Extra tooling pins are required around the base of each progressed section of FPW to resist the tendency for the layers to shift outwardly—i.e., at the A and B board edges in the example. Unless enough pins are used, the longer FPW layers will attempt to thrust into the rigid area during lamination, distorting nearby circuitry locations.

- If the cap boards are windowed to allow the FPW to bulge outward, the lamination fixture must be relieved to accommodate the FPW bulge. Extra plates may be required to build up the fixture thickness so that the bulge can't touch the press pads.

- At numerical control (NC) drilling and routing the height of the presser foot and tool must be set so that they clear the top of the bulge. It may be necessary to invert the RF panel over a support plate which is pocketed to accommodate the bookbinder bulge and machine from this flat side; drill and rout programs are mirrored in this case.

- Resist lamination may require additional filler plates or shims so that adequate pressure is applied to the photoresist film. Sheet lamination with vacuum assist is mandatory.

- Bookbinder results in lower installed stress but higher stress in the flat condition. Throughout finishing or assembly stages, a bookbinder RF panel has unusual stresses in the arching FPW layers. Baking, mechanical shocks, or flexure may weaken the bond at the junction of FPW and rigid areas or fatigue the foils.

- Beading—application of semihard compounds in a fillet along the root of FPW layers to distribute bending stresses—may be much harder to apply in bookbinder. See Fig. 4-3 and Figs. 9-11 and 9-12 for illustrations of progression/bookbinder design.

Machining

NC-controlled slot routing and steel rule die windowing are performed with reference to tooling holes, which are used in abundance in RF.

Rigid areas, because they contain glass-fabric-reinforced layers, can't be die cut; they must be routed to create smooth, unstressed, accurately contoured edges. FPW doesn't rout gracefully, particularly where it isn't tightly gripped—for example, in unbonded flexing areas. One way to deal with this problem is to preoutline flex-area FPW edges at the individual layer stage by die slitting (leaving the scrap in place) before multilayer stack-up. After all wet process steps, including tin-lead reflow (if used) and cleaning, are completed, the panel returns to the router for final outline.

Cap board slots are already there, extending beyond the edges of the preslit FWB layers beneath them. The router bit is plunged through the panel at the edge of a slot and progressed around the rigid area contours until it rejoins the slot—or encounters another slot—and so forth, to generate the rigid area outline. Routing also cuts into the edges of the flex areas, but only enough to allow the circuit to be depaneled; the precision precut of the FPW edges defines them.

When the RF circuit is lifted out of the routed panel, unbonded areas of cap board lying above the pouch (flex) areas remain attached to the rigid areas by the bag layer, and a completely outlined circuit—except for debagging—is achieved. The thin hinge of bag holding the cap board scrap is simply sliced off to free it; the tiny piece of bag which remains acts as a strain relief/smoothener to protect the FPW layers from the sharp edges of the cap board. Figure 9-14 is a view of an outlined, debagged RF circuit. A cap board slot is visible, extending into the scrap area to each side of the final outline rout; bag film is visible extending from the edge of the round cap board area.

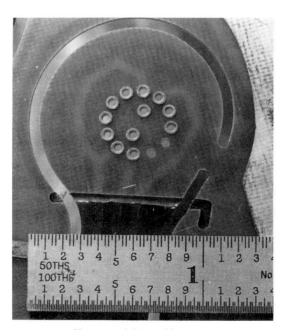

Figure 9-14 Close-up of slot and bag.

Figures 9-15 and 9-16 show an outlined and debagged RF circuit. Figure 9-15 is an overall view which illustrates use of scrap edge strips to hold the rigid areas together during final assembly, protecting the FPW interconnections from damage. Figure 9-16 is a close-up of the breakaway feature—a series of small holes on 0.05-in centers— plus silk-screen printed nomenclature which identifies the connectors.

Figures 9-17 and 9-18 show another palletized RF panel, a two-up circuit. Although the panel has not been assembled at this point, flex

Figure 9-15 Overall view of palletized RF circuit.

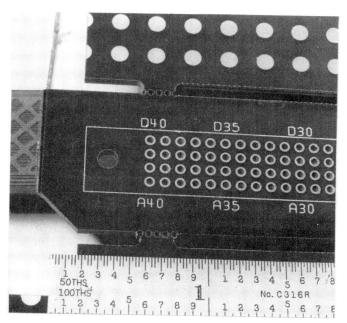

Figure 9-16 Close-up of breakaway feature and screen-printed nomenclature.

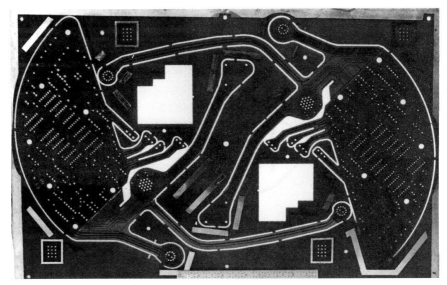

Figure 9-17 Bagpipe panel.

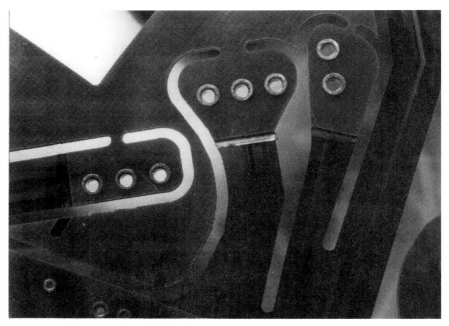

Figure 9-18 Close-up of bagpipe heads.

areas have already been debagged to allow inspection of the FPW. Four process-control coupons are visible. Figure 9-18 is a close-up of three small PTH-terminated rigid areas showing the bag flap, cap board slots, and breakaway feature. Etched-copper tear stops are also visible.

Plasma

Plasma treatment of RF is similar to treatment of multilayer FPW with the exception of pouch venting and unique etchback problems caused by adhesiveless materials. Plasma attacks organics at variable rates. Conductors and pads in conventional FPW are surrounded by soft, readily removed adhesive, thus plasma treatment creates an opening on both the top and bottom of the PTH pads for strong, *three-point lock* (top, edge, bottom) electroless copper attachment. Pads in adhesiveless FPW are supported directly on polyimide film, a tough, plasma-resistant polymer. Consequently, in adhesiveless cross sections it's expected that etchback will be possible only on the coverlayer side of the conductors—into the coverlayer-attaching adhesive or interlayer bondply.

RF panels built with prepreg adhesive layers require both adhesive removal and glass-fiber etch for good electroless adhesion. Uniform surface treatment and/or removal of glass fibers is difficult to achieve, and variable penetration of electroless copper into fiber bundles should be expected.

Figures 9-3 to 9-8 illustrate etchback and fiber conditions in conventional and adhesiveless RF constructions.

Surface preparation. When polyimide films are present in RF construction, surface preparation for lamination is mandatory. Polyimide films are difficult to bond because their surfaces are smooth and chemically resistant. Four methods are used:

- Plasma treatment in O_2 (oxygen wash)
- Pumice scrub
- Acetone wipe
- Preexisting adhesive coating

Pumice scrubbing roughens and dulls the shiny surface to improve mechanical bonding but is aesthetically undesirable if used on visible areas and can be cause for rejection. As is always the case with pumicing, thorough rinsing to remove all abrasive residues is important.

Acetone wiping is somewhat hazardous but quick and effective; it should be done within minutes of layup. But watch out for water pickup in the solvent—acetone is hygroscopic. Keep it covered.

Oxygen wash is attractive because it desiccates the layers and avoids explosion hazards and foreign material risk while improving adhesion. It usually requires racking, a labor-intensive, troublesome process, because the interior of a plasma treater does have gas flows, and sheets or layers of etched details must be secured or they'll swing about and stick to each other, interfering with uniform treatment.

Use of adhesive double-clad FPW laminate gets the benefit of adhesive precoat without added expense or adhesive; exposed dielectric areas between the conductors in the etched patterns on both sides of the laminate are readily bonded with coverlayer or bondply.

Any of the treatments should be performed shortly before lamination (within 8 hours); the advantage of buying adhesive precoated material is that the benefit is permanent.

Surface treatment of the etched-pattern side of a laminate isn't necessary. Pretreatment of the polyimide film which is used to gain good metal adhesion at the laminate stage, including adhesive coatings, persists through etching, and adhesion to the conductors and pads (which should be oxide-treated) reinforces adhesion to the dielectric.

A good test for adherability is to use surface-energy measurement solutions. Bondable surfaces will test 70 dyn/cm^2. Ordinary water can be used for testing—when it wets onto the polyimide, adherability with RF and FPW adhesives will be adequate.

Materials

RF circuits are made in virtually any number of layers. Yield drops exponentially with increasing layer count because of increasing opportunity for error and difficulty in maintaining alignment and managing material properties through lamination, drilling, and the PTH process. As more stable materials are used, and manufacturing precision and process control improve, the feasibility of building high-layer-count RF also improves. A general statement could be that RF designs with up to 18 layers are buildable at reasonable cost.

A key factor is layer-to-layer alignment, which must be within 0.014 in in MIL-spec products. Meeting this requirement takes material with good stability, small panels with plenty of tooling pins, and lots of stabilizing copper in the layer designs.

Vigilant inspection and control is required for all materials. Reactives such as prepreg, cast film adhesive, and coverlayers must be stored properly, periodically tested and discarded when performance drops or shelf life is exceeded. Materials contribute very little to RF cost: it's very poor economy to use any but the best and freshest. Materials are an important factor in FPW production; they're more important in the production of multilayer FPW and critically important in RF circuit production.

Cap boards

Cap boards are the armor which protects RF circuitry from assembly stress. A frequent choice is polyimide-glass, which easily withstands mass soldering and rework cycles, if necessary. Where initial product cost is particularly important, epoxy-glass or other materials are used. Cosmetic issues such as weave exposure, scratching, and chipping are sometimes significant; polyimides are more easily damaged.

FPW materials

Many RF circuits have been built with ordinary FPW material stackups including acrylic bondplies and cast adhesives. These circuits are usually polyimide-based, but other dielectrics such as FEP Teflon, aramid papers, and flexibilized epoxy-mat systems have also been successfully used. Conventional acrylic and modified epoxy systems are least expensive to buy and process and are satisfactory for low-layer-count, thin-foil circuitry. Prebaking laminate to relieve built-in stress is used by some manufacturers as added assurance of material quality; if the baking temperature is high enough, this process also weeds out poorly bonded material. Layers are etched and fully coverlayered just like conventional FPW, then built into multilayer panels for RF processing.

When the layer count rises, when heavier foils (above 0.0014 in) are required and MIL-spec qualification is needed, systems other than conventional flexiblized FPW materials are more cost-effective. As discussed in Chap. 5, "Dielectric Materials," and shown in the thermomechanical analysis (TMA) curves of Figs. 9-20, 9-21, and 9-22, conventional FPW adhesives have enormous expansion at elevated temperatures (higher than 250°C). If acceptance includes thermal stress test, which is applied only to the plated throughhole (PTH) (rigid) region, it's good practice to exclude highly plasticized, low cross-link density adhesives from those areas. This is accomplished by a variety of construction techniques (some patented) which include:

- *Prepreg bondplies.* FPW layers are conventionally constructed and coverlayered with FPW material throughout but bonded together in rigid areas with prepreg (usually similar to the cap board system—epoxy with epoxy, etc.). FPW adhesive is present in both base and coverlayer but removed from bondply/cast-film layers.

- *Inlay processes (bikini coverlayer).* Both the coverlayer film and adhesive are deleted from PTH areas of the FPW layers by windowing and replacement with prepreg so that the flex areas have conventional coverlayers and rigid areas have prepreg. FPW adhesive remains in the base laminate layer.
- *Custom lamination.* All FPW adhesive is taken out of the rigid, PTH areas. This requires preparation and use of a special laminate, which has windowed, inlaid adhesives bonding foil to film—flexible, plasticized adhesive in flexing areas and prepreg or other stiff adhesive in rigid areas. See Fig. 9-13. This process, combined with inlay coverlayering, yields PTH areas that contain only rigid dielectric materials and polyimide films. Note that the custom laminate process take more labor, requires special laminate, is more susceptible to foreign material (which enters during the production of the inlaid base laminate), and has reduced dimensional stability compared to standard sheet or roll-made laminates. Location of rigid and flexible areas must be accurately related to tooling holes in the laminate so that the circuit image can be correctly registered to the custom dielectric structure.
- *Adhesiveless laminates.* FPW layers built in adhesiveless materials eliminate the need for custom lamination or inlaying and associated cost and quality concerns. More stable than conventional FPW materials, rigid-like in glass transition temperature T_g and coefficient of thermal expansion (CTE), these materials allow production of high-layer-count, high-density RF circuits with greater yield and production predictability.

Adhesiveless FPW with prepreg bonding layers and bikini coverlayer is shown in Fig. 9-19. In this sketch the left side represents a rigid area; the right side, a flexible area. Two of the six double-clad FPW layers are shown, with the junction between 0.001-in Kapton*/0.002-in adhesive FPW coverlayer and corresponding 106/1080 prepreg in the rigid area aligning with the inner edge of the slot. Slots in the cap boards are shown, aligning with the rigid-flexible interface. The bag layer is the two plies of 106 prepreg which lie between the filler and inner cap board layer. These prepreg layers are scored—glass fibers are cut—at the slot, so that after lamination the scrap part of the cap board, consisting of the five conductor layers (foil layer M24 plus etched layers M20, M21, M22, and M23, for example) with the associated cores and seven plies of prepreg, can be broken across the scored slot and removed.

---

*Registered trademark of DuPont.

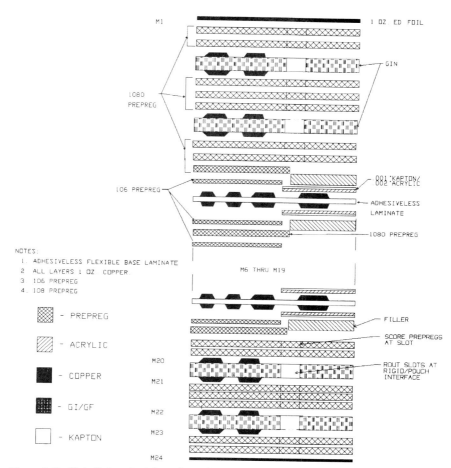

Figure 9-19 Detailed material stack-up for adhesiveless-FPW-based RF construction.

Adhesives, thickness

Good lamination requires uniform thickness, because laminating pressure distributes itself according to thickness: higher pressure on thicker areas and lesser on thinner areas, potentially resulting in a poor bond. In RF processing, unlike FPW, flat, smooth outer surfaces are needed because imaging comes after lamination—it's difficult to get tight resist contact with irregular cap board surfaces. Flat surfaces also aid drill accuracy. Conformation of coverlayer around an etched run aids FPW encapsulation, but can't be used in RF. Consequently, two important differences in adhesive usage between FPW and RF emerge:

- More adhesive is required in RF to provide internal accommodation for etched features.
- Where the adhesive is windowed away to provide unbonded areas, fillers are inserted to shim the stack-up to uniform thickness, control bonding, and restrict adhesive flow.

Conventional FPW adhesives in either bondply or cast-adhesive form can be used for RF production, and are cost-effective in low-layer, simple designs. More complicated or stringently tested products require better adhesives. FPW adhesives have low flow to help keep coverlayer apertures open and free of adhesive intrusion in bare-pad designs. Low flow is undesirable in RF constructions because it forces use of greater adhesive thicknesses for etched feature encapsulation. Conformation of coverlayer around etched runs aids in low-flow adhesive encapsulation in FPW, but can't be used in RF—each layer must have enough adhesive to smoothly embed etched features without telegraphing into the next layer. Furthermore, plasticized FPW adhesives have undesirable memory—they resist displacement during lamination because they're not completely thermosetting, and always attempt to return to their original film thickness when heated.

PWB prepreg resins are much better suited to multilayer lamination, because they become very fluid during the initial phases of cure, flowing around and encapsulating the etched features. When the cure progresses into cross-linkage, the resins lock in position but without residual stress—they have no tendency to return to their initial sheet form. A wide range of resin compositions, glass percentages, and flow properties is available in prepreg form to ease choice of the right system.

Adhesive CTE. One of the first properties to be advertised and discussed in FPW laminates is peel strength. However, peel strength is an uncertain and nonrepresentative property in flexible laminates—hard to measure, variable, not related to normal failure modes. It has the single good feature of being numeric and therefore easily comparable.

High peel strength is a result of adhesive extensibility or elasticity; stronger (but rigid) adhesives have lower peel values because their nature leads to stress concentrations in the peel mode. The unfortunate fixation on high peel strength in the FPW industry has led to the development, specification, and almost exclusive use of flexibilized adhesives which have high peel values. Unfortunately, flexibilization also leads to large CTEs at temperatures above T_g. Unimportant in low-layer-count constructions, large CTE is devastating in multilayer or rigid-flex circuitry.

The ideal FPW laminate CTE matches copper at 17 ppm/°C across a range from below 20°C to soldering temperatures of 270°C or better. When given at all, CTEs are only cited for ambient conditions (below T_g).

Figure 9-20 is a thermomechanical analysis curve for a popular acrylic adhesive. The sample consisted of several layers of cast-film adhesive, bonded together and cured to create a test specimen 0.6960 mm thick to provide greater measurement accuracy. It was baked for 16 hours at 125°C to simulate standard RF drying and stabilizing processes, then jacketed in aluminum foil to prevent probe penetration. In the test procedure it was first subjected to a settling run at 100-g load at 100°C for 5 minutes, then cooled to −20°C and finally exposed to rapidly rising temperatures (10°C/min) while the Z expansion was continuously charted. The CTE is the slope of the curve; for this sample, the total expansion was 0.15 mm, and the average expansion rate was 21% or 860 ppm/°C.

Expansion of a second sample, a modified epoxy, is shown in Fig. 9-21. Prebake and settling were identical; expansion in a 0.3556-mm sample was 0.042 mm for 12% average expansion rate or 480 ppm/°C.

This adhesive system, because it is more cross-linked, has much less expansion—slightly more than half—than the popular, higher-peel acrylic system. Because it's more cross-linked, this adhesive yields lower peel test strengths and has been withdrawn from the market.

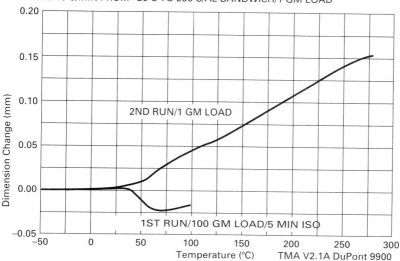

Figure 9-20 Thermomechanical analysis plot of expansion versus temperature for a typical acrylic FPW adhesive. Expansion to 275°C is calculated by dividing the expansion by the sample thickness; in this case, (0.15 mm)/(0.6960 mm) = 22%.

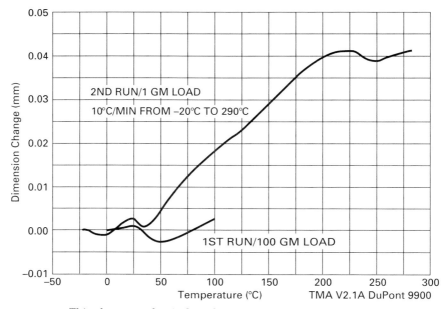

Figure 9-21 This thermomechanical analysis plot shows reduced expansion of more highly cross-linked epoxy adhesive over the MIL-P-50-884 temperature range. This expansion is (0.042 mm)/(0.3556 mm) = 12%.

Adhesive is only one of several components in an RF construction. To test the effect of flexibilized adhesives in RF circuits, expansion for three panel constructions consisting of cap boards and FPW layers was measured and is shown in Fig. 9-22. In this graphical presentation the vertical axis is expansion; horizontal is temperature.

Coupons were cut from a typical acrylic-bonded 10-layer MIL-P-50884 qualification board, a similar board built with adhesiveless FPW layers and polyimide prepreg bonding layers, and a doubled-up version with twice the number of FPW layers (increased from 8 to 16) in adhesiveless construction with prepreg bonding layers. The test produced these results:

- The acrylic sample literally ran the test instrument out of headspace at 223°C; to this point the expansion from room temperature averaged 281 ppm/°C.

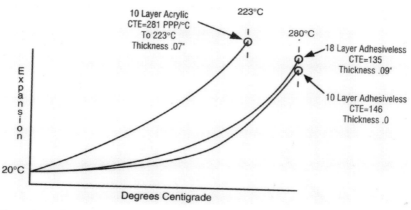

Figure 9-22 A comparison of the expansion of typical RF constructions, one with acrylic adhesive and the other adhesiveless. The comparison shows the substantial reduction in expansion that can be achieved by eliminating materials having low cross-link density.

- The 10-layer adhesiveless and prepreg construction coupon was much more stable; CTE average to MIL-spec 280°C is 146 ppm/°C.
- The 18-layer sample, with twice the number of FPW layers, expanded slightly less, averaging 135 ppm/°C. Both adhesiveless-prepreg constructions survive thermal stress and thermal shock test with no evidence of failure.

It's important to remember that an RF panel is locked together by plated copper barrels and pads. CTE for copper is around 17 ppm/°C. Expansion of the dielectric system at a greater rate than this stresses the barrels. When the difference is large enough, failure occurs: barrels fracture, pads lift, or the dielectric tears itself apart.

An ideal RF dielectric system matches copper CTE; the system consists of PWB layers in the cap boards and polyimide films and adhesives in the FPW layers. It's known that conventional FPW adhesives have enormous CTEs compared to copper; the purpose of this test was to determine the contribution of conventional and adhesiveless FPW systems in practical RF constructions. As the results show, adhesiveless FPW systems with prepreg bonding layers produced composite expansions which are less than half that of conventional FPW systems.

Barrel fracture can result from excessive expansion at thermal stress or from repeated-cycle fatigue. MIL-P-50884 requires ability to withstand a minimum of 100 thermal shock cycles (fast cycling) from −65°C to +125°C. The 18-layer adhesiveless-prepreg construction withstood over 200 cycles of this test without loss of continuity or cracking.

Prepregs

Prepregs are manufactured in a wide range of resin contents and flow properties. For RF use, high-resin-content, fine-weave, low-flow prepregs are preferred; examples are 60 to 65% resin in 106 or 1080 cloth with 4 to 6% resin flow. Materials of this description will accommodate roughly 50% of their thickness; i.e., 0.005 in of prepreg will encapsulate 0.0025-in etched features; 0.005 in of prepreg (roughly three sheets) will embed two facing 0.0014-in etched layers with sufficiently easy flow to leave the base polyimide films flat. In other words, distortion (conformation) of the FPW layer is not required; there's no telegraphing of etched pattern from one layer to another.

Acrylic adhesive can accommodate about one-third of its thickness, but, unlike prepregs and other thermosetting polymers, always retains some thermoplastic tendency to spring back to flat film form, a tendency which contributes to failure at thermal stress testing.

Flow control. High adhesive flow is desirable for efficient encapsulation (defined as complete fill with minimum adhesive thickness), but undesirable into pouch areas. Remember that FPW layers are heavily windowed and precut to facilitate machining to final contours. Picture, in cross section, the junction of a rigid area and adjacent unbonded flex area. The rigid area has multiple plies of prepreg; at the corresponding point in the flex area there are fillers between layers to prevent bonding. When the panel is heated and compressed during lamination, the prepreg resin liquefies and flows until it gels. Flow of resin into the spaces between and around FPW layers is undesirable because it's very difficult to remove without damage, forcing a choice between the benefits of high flow during lamination and the need for maintaining a clean edge.

The problem is addressed along two paths:

- Fillers are chosen to be the correct thickness so that they fill all the spaces between FPW layers, leaving the least possible space for resin flow.
- Prepregs are tightly specified and tested to assure correct resin content, flow, and gel time (duration of fluidity at cure temperature). Consistent with these controls, lamination conditions (rate of temperature rise, pressure, cure temperature) are also controlled.

Prepregs can absorb moisture in storage, which interferes with flow and gel, consequently desiccation by storage in dry gas or maintenance in a vacuum press or turkey bag, under vacuum, for several hours prior to lamination, is recommended.

Alternative constructions

Until the widespread use of use of adhesiveless FPW materials and prepreg bonding layers, construction of RF circuitry was a low-yield, high-cost business. Use of RF circuitry and design of new circuits continued, however, because of the unquestioned benefits of RF in package efficiency and reliability. An alternative construction method, aimed at reducing the procurement cost while retaining most of the benefits, has been proposed. This is a modular technique—a return to a concept of separate FPW and PWB circuits, assembled to form RF composites.

Advantages are that the multilayer, PTH-interconnected rigid areas can be fabricated by PWB techniques and therefore are considerably less costly and more available. The FPW interconnectors can be simple jumper circuits of parallel runs with terminating hardware or may be custom-configured, if necessary. In the modular technique, should either FPW or PWB be damaged, separate replacement is possible, saving some of the cost of the circuit (and assembly, if hardware is already installed). Overall cost of a built-up circuit made by using the modular approach is estimated to be 30% less than the cost of an equivalent RF circuit.

Disadvantages are that each FPW conductor requires two added solder (or pressure) connections—one at each end—compared with an integrated RF circuit. Added connections, compared to the coextensive conductor system in RF circuitry, cost money, require two added pads per conductor (space for which normally isn't available in RF designs), and introduce reliability concerns.

Photographs

The stack-up of adhesive in a typical RF panel will amount to roughly 50% of thickness. Given the huge thermal expansion of FPW adhesives, trapped within rigid cap boards that are riveted together by PTH barrels, the result is destructive pressure buildup inside an RF panel. Copper barrels expand at 17 ppm/°C; FPW adhesive expansion (see Fig. 9-22) can be over 800 ppm/°C.

It has been calculated that the internal pressure buildup in an RF coupon with rigid cap boards reaches 30,000 psi or more. This extreme pressure crushes the PTH barrels inwardly, potentially separating them from the pads and causing inner land disconnect; it forces cap boards outward in a bulging arc, from which they may not completely return as the coupon cools. The outward-bulge, incomplete-return theory could explain why laminate voids, which are caused by excessive pressure, look like bubbles formed by inadequate pressure: the internal compression in the coupon may actually be negative as a result of cap board displacement.

Figures 9-3 and 9-4 show acrylic-bonded coupons after thermal stress; both show laminate voids (tears or bubbles) in the acrylic adhesive layers. These pictures also show the high percentage of cross section occupied by adhesive (for identification purposes, adhesive surrounds each rectangular, whitish conductor); distorted/rippled polyimide-film layers which are a consequence of low adhesive flow; resin-glass cloth cap boards; and etchback at barrel-pad interfaces.

Figure 9-23 is a section from the manufacturing drawings for the product shown in Fig. 9-4. Total FPW adhesive thickness allowed is 0.022 in in a total thickness of 0.055 to 0.069 in. This was an essentially unbuildable product in conventional FPW materials because of

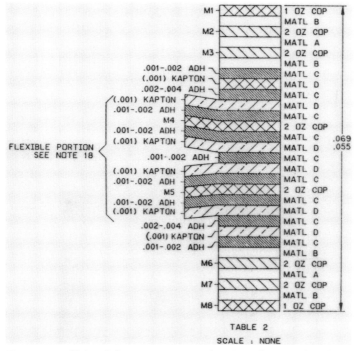

Figure 9-23 Extract from a customer drawing showing details of material call-outs for RF constructions. This is an eight-layer circuit with four 2-oz-copper layers.

inadequate adhesive to accommodate the 0.0028-in conductor thickness and excessive adhesive for thermal stress.

For contrast, Figs. 9-5 and 9-6 show accepted prepreg-bonded coupons. The central polyimide films in the flex layers remain flat

Figure 9-24 Cross section of an 18-layer RF circuit after thermal stress.

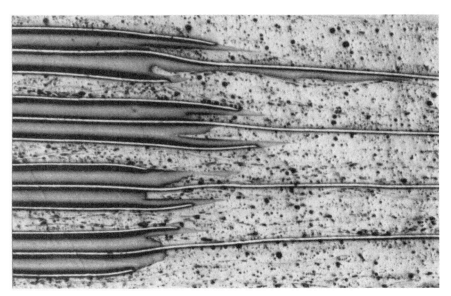

Figure 9-25 Cross section of the coverlayer-rigid interface of the same circuit shown in Fig. 9-24.

and undistorted as a result of adequate resin flow; there are no voids or evidence of dielectric failure because the dielectric has low CTE and high T_g.

Figure 9-7 is a close-up of the pad/barrel interface in a direct-metallized adhesiveless construction of 0.0014-in foil on 0.002-in polyimide film. This is a coverlayered circuit; the coverlayer is bonded onto the pads by layers of prepreg (note glass bundles). The dielectric is unaffected by thermal stress. The degree of etchback is inhibited by the absence of an adhesive layer between the foil and polyimide film. The picture shows the horizontally oriented fine-grain structure of the high-performance electrodeposited metallization. The excellent etching properties of this foil are shown by the near-vertical edges.

Figure 9-8 is a similar view. This is an example of custom-built laminate with prepreg bonding foil to polyimide film. Etchback is more uniform; rolled foil shows fine grain with tendency for horizontal arrangement; etch properties evidenced by curving edges of 0.0028-in foil on 0.001-in polyimide film.

Figures 9-24 and 9-25 are, respectively, a thermally stressed coupon from an 18-layer RF circuit built on directly metallized FPW layers and a close-up of the FPW-rigid area interface of the same board. Note in Fig. 9-24 the flat, undistorted FPW layers and solid, undamaged prepreg dielectric system. Figure 9-25 shows the smooth, void-free intermixing of flexiblized acrylic adhesive in the coverlayered FPW circuits with the rigid polyimide resin in the rigid section. Figure 9-22 shows the expansion performance of this construction.

Summary

Multilayer FPW with rigid layers at termination areas—RF circuits—form the most comprehensive, reliable interconnection technique for high-density packages.

New manufacturing concerns include complicated outlining; added tooling for windowing and filling internal adhesive layers; venting and resealing internal nonbonded areas, and materials engineering to meet MIL-specification-required thermal stress and shock testing.

Adhesiveless materials and high-flow, highly cross-linked adhesives aid production of reliable, thermally stable dielectric structures which survive severe testing with high yield.

Chapter 10

Standards and Specifications

Introduction

Flexible printed wiring (FPW) is inspected and tested out of all proportion to its cost. There are many reasons for this extreme scrutiny, beyond the obvious fact that FPW has both mechanical and electrical properties:

1. As the foundation of high-value assemblies, FPW's quality and reliability are critically important—far more important than suggested by its relatively low price. FPW is like the proverbial horseshoe nail, for want of which the battle was lost: inexpensive compared with assembly cost, but if it's lost, so is the assembly.
2. Because FPW is custom-manufactured, each design is unique. Uniqueness inspires suspicion and increased inspection.
3. Earliest FPW use was in military electronics. The associated abundance of specifications and standards continues.
4. A wide range of materials and process technology is involved, thus there are many failure modes.
5. Transparency cries out for inspection.
6. A universal quality-control attitude is: when in doubt, specify.

As complete understanding of FPW manufacturing technology isn't needed for procurement, many FPW buyers use blanket tolerancing and more specifications than required, simply because they don't know the requirements that are necessary and sufficient to control product quality.

Overspecification goes with the current trend to buy, rather than make. When a design is intended for outside procurement, fear of the unknown leads to overspecification. When products are produced in-house by one department for use in another, rationalized quality standards are used—in this situation, nobody gains through excessive specifications.

The near certainty of excessive requirements should be anticipated and fought by managers and designers. FPW is a flexible, formable product which performs an electrical function. To a first approximation, if it fits the hardware and provides all the needed interconnections with no unwanted ones, it's successful.

Cost Impact of Quality Requirements

A critical goal in every procurement negotiation is the establishment of rational inspection requirements because cost is directly linked to quality. FPW is custom-manufactured: each production run is a new event and every design has unique requirements. Yield is a particularly important element in FPW cost. Percent profit and variation in yield in FPW manufacture are nearly identical—a bad day at the inspection station can absorb a week's profit. Manufacturers calculate FPW prices to offset yield losses based on past experience as a result of inspection criteria and add warranty reserves and contingency offsets to that figure.

Sources

Specifications come from three sources: industry standards, customer requirements, and contractual agreements.

Industry standards

Industry standards are established with significant volume and use, when familiarity and experience establish practical levels of quality as competition between sources sets reasonable prices. Simplest cost/quality fix is the *catalog item buy* because designing to use a catalog item and ordering by part number invokes manufacturer's specifications and thus assured performance, cost, and quality. Unfortunately, except in rare instances exemplified by jumpers and parallel conductor interconnect cables, no FPW design is sold in enough volume to reach this stage.

Customer standards

When one customer buys enough product to become a significant factor, it can set its own standards and require compliance. That's the situation with military specifications: a large percentage of FPW production goes into military electronics—has from the beginning—thus there's a complete dossier of standards defining FPW from the military customer's point of view. These standards tend to be unnecessarily strict and therefore raise product cost, but they've been used successfully for many years, are widely understood and accepted, and have no hidden faults.

Contractual agreements

It's known that improvement and change occur often in electronics and widely recognized that new technologies aren't adequately anticipated by established specifications. When different materials or manufacturing methods are proposed or needed to satisfy an urgent need, deviation from existing standards or development of ad hoc quality requirements is required to arrive at preaward agreement.

Wisdom suggests that widest possible use of established standards should be the rule, modified by clear definition of required product performance. The negotiation between designer and potential manufacturer is a potent stimulator of design changes that preserve performance but improve manufacturability. The give and take between what's wanted and what's possible stimulates good thinking and creativity leading to workable designs and reasonable prices.

Time spent developing a good set of specifications can be repaid many times over in smooth, uninterrupted product delivery. Any mutually agreed set of standards may be used, ad hoc, industry-derived or customer-based, but experience indicates that well-used and tested standards such as those produced by the Institute for Interconnecting and Packaging Electronic Circuits (IPC) or military services provide less opportunity for misinterpretation and grief. Quality requirements and referenced documents should be listed on the purchase order. The order of precedence should also be stated, particularly with reference to artwork, which frequently conflicts with dimensional requirements.

The combination of data package and quality requirements expresses the buyer's expectations to the manufacturer. Experience proves that most data packages have errors or ambiguities. These have to be resolved at some point: better to do it before production and the procurement clock start. Clarity, accuracy, and simplicity seem to be unreachable goals, but the closer the procurement documents come to this ideal, the sooner production can begin and the more likely finished parts will satisfy the buyer.

Double Dimensioning

Double dimensioning is common in FPW designs. It creeps in because databases include dimensioned drawings and artwork, which is also a dimension-controlling document. To further complicate the situation, computer-aided design (CAD) artwork data may include allowances for etch loss and material shrinkage which are incorrect for the chosen vendor.

IPC standards clearly state that the FPW manufacturer is responsible for verification to assure that artwork is compatible with drawings and inspection requirements. Regardless of whose specifications are invoked, it's wise to place responsibility solidly on one party or the other. Artwork, in effect, is a controlling and inspection document and must be very carefully reviewed to assure that it produces FPW which matches the required dimensions. The buyer should strongly urge review and cross-checking of the data package and should not be surprised if unexpected conflicts are discovered in packages which have already been used for procurement. It is not unknown for conflict to have existed and somehow been overlooked for several procurement cycles before a clash occurred.

Document Systems

Systems of specifications and standards for FPW, rigid-flex (RF) circuitry, and printed wiring board (PWB) are available from:

1. Institute for Interconnecting and Packaging Electronic Circuits (IPC), 7380 N. Lincoln Avenue, Lincolnwood, Illinois 60646.
2. The military services. A commercial reprinting service is Document Center, 1504 Industrial Highway, Unit 9, Belmont, California 94002.
3. Underwriters Laboratories, Inc., 333 Pfingsten Road, Northbrook, Illinois 60062.
4. Canadian Standards Association (CSA), 178 Rexdale Boulevard, Toronto, Canada M9W 1R3.

The IPC, a printed circuit industry association founded in 1957 which today has more than 1800 worldwide members, offers nearly 200 test method, materials, product, and quality systems documents covering every aspect of FPW, RF, and PWB. Coverage of FPW and PWB technology in IPC publications and standards is comprehensive, constantly updated, and expanded through twice-yearly meetings.

In addition to quality documents of all sorts, the IPC has an extensive library of videotapes of PWB and FPW manufacturing technology as well as technical papers and test reports which are available to be studied for specifics.

Specifications and standards from the IPC or military services provide an objective, third-party basis for negotiation between buyer and seller. These established, familiar libraries are easy to invoke and have the enormous advantage that convention and practice help in the event that ambiguity or dispute arises. Most of these standards are structured around multiple levels or classes of quality. This allows universal procurement use, avoiding confusion brought on by unknown standards, with specific quality levels invoked for each purchase contract according to the design and intended use.

Guidelines

Avoid subjective standards such as "good workmanship" or "high quality"; prefer numeric and objective over visual standards. Include test coupons in the composite layout. Use these for destructive tests such as cross-sectioning or peel tests, or retail them as samples for rework simulation, infrared (IR) irradiation, or cleanliness verification.

Mechanical

Definitely undesirable are tight tolerances—anything less than 0.01 in on mechanical dimensions other than thicknesses or hole to hole within a cluster.

It's not practical to measure long lengths of FPW. The product simply refuses to lay down flat and straight; measurement means holding under tension or flattening under glass plates, procedures which take time and compromise accuracy. In addition, because FPW is flexible, it's always designed to be somewhat longer than the theoretical length to allow for slight bends and curves and for installation and service which makes close tolerances irrelevant and unnecessary. Fractional tolerances for long lengths are easy to measure, appropriate to the required precision, and save money and aggravation.

Within clusters where FPW must align with hardware pins, machine tolerances of 0.002 in are producible, measurable, and appropriate.

Material call-outs can be shown in cross-sectional views and/or drawing notes with preferred materials and thicknesses. It's wise to leave as much flexibility in construction as possible.

Forming should not be tightly defined, because it's very difficult to measure a bent FPW and almost impossible to maintain a desired shape without jigs. Show location of bend lines by reference to artwork features or tooling holes.

Location of stiffeners should also be referenced to tooling holes or artwork hash marks. Method of attachment—adhesive or hardware—can be specified on drawings or in notes. Define stiffener material by drawing note or call-out.

Give a total count of holes (by diameter). The number of conductor layers or any other repeating items should be given. It's surprising how often a cross-check of this kind reveals data-package errors.

If external nomenclature is required, the letter size, ink type and color, and preferred location must be specified.

Conductor surface finish [OPC (organic protective coat), solder, other plated finish] must be defined.

Electrical

Electrical requirements are numeric and consequently easy to specify. When the resistivity of a conductor run is important, the appropriate quality control vehicle is a drawing note which specifies maximum resistance between terminals. Visual inspection for line width, notches, "mousebites," and scratches cannot duplicate a resistance measurement, since resistance is an averaged property—severe necking won't significantly affect overall resistance (see the Chap. 4 section, "Connector Approaches").

Watch out for hidden double dimensioning. Resistivity is a property of matter: if conductor material and thickness are specified, and line width and spacing are also specified, the resistance is already determined. Be certain that the artwork provides enough conductor area to satisfy resistance requirements with reasonable manufacturing allowance. Requirements for impedance or capacitance to ground are other possible traps where design conflicts with material properties.

Electrical properties such as dielectric strength, insulation resistance, capacitance, and impedance are measurable and can be defined by drawing notes.

Visual

Visual inspection for any characteristic should have specified magnification limits. The inherent transparency of FPW seems to invite visual inspection at every stage, even after shipment and delivery to the customer's inventory. The subjective and intellectual appeal of microscopic examination frequently leads to excessive time spent studying defects. Limits of magnification should be set; a reasonable and common value is $10\times$.

Most foreign inclusions in FPW are organics which, although aesthetically displeasing, present little risk of long-term degradation. When foreign material is detected, there should be a referee process for determining acceptability. A suggested and sometimes used policy is to apply a specified overvoltage to adjacent conductors or to the overlying insulation layers. Breakdown or insulation resistance less than some multiple of the basic requirement can be used to define rejection.

Summary

Comprehensive specification packages controlling materials, process, and product are available from the military services and the IPC.

Specification choice has profound impact on product cost. Reasonable requirements lead to lower cost and adequate quality with reduced inspection labor and MRB activity.

Double dimensioning is a recurring problem in FPW designs, with conflicts typically occurring between artwork data and dimensional drawings or electrical requirements.

Chapter 11

Assembly

Background

When FPW appeared in the mid-'50s, the electronics industry welcomed the innovation but shied away from using FPW because there weren't any specifications for the product or assembling a package with it. Specifications and standards provide stability and safety but they do so by suppressing innovation. They're anchors to the past, locking in the familiar and safe.

Today, solder attachment is the main assembly technique used with FPW, at least partly because there is a complete dossier of established methods and inspection standards as well as a large cadre of trained, licensed practitioners of hand-solderers. But in the early days, assembly was a seat-of-the-pants, make-it-up-as-you-go process. In this uncertain situation, users felt they were risking their careers by using FPW as the interconnector. Standards and specifications defining good technique and acceptable joints hadn't been written; wire was the safe choice.

At that time, hardware was sold with wire-type terminals—hooks, eyes, spades—in a wide variety of shapes, but all intended for wire. When such components were needed for an FPW-based harness, the first step was to adapt the terminal pins to engage an FPW pad.

NASA and the frenzied space programs of the '60s did more to boost FPW acceptability than any other single factor. FPW was needed in space for technology reasons: it increased the functionality of instrument packages while reducing weight and improving duplication of a prototype to a flight unit. Thus motivated, NASA quickly evolved standards for joint acceptability; NASA trained inspectors who traveled across the country with standardized rules and requirements; NASA developed soldering practices, trained solderers, and administered a licensing and certification program.

With space use came publicity and wider familiarity. Gradually, FPW moved from the curiosity stage to the somewhat acceptable stage. If NASA said that FPW assemblies were flightworthy, then similar technology could be used for computers or telecommunications—certainly in cameras and the like. Thus, coasting forward from the big space-use boost, FPW infiltrated the packaging engineer's tool kit, gradually gaining respectability. Constant missionary work by FPW builders plus passage of time without serious mishap has brought, at last, general acceptability. FPW has shortcomings, but overall is accepted as a solidly reliable, cost-saving, performance-boosting—if initially hard to procure—platform on which to build electronic devices.

Joining Processes

Solder

Solder attachment is the most common joining method for FPW. Hand-soldering techniques are applied to almost any sort of joint design; automated methods of wave, elevator/dip, infrared (IR) and hot-air or vapor-phase reflow deposition are used when volume justifies tool development.

Preparation for soldering takes these steps:

- Polyimide dielectrics (or any system which absorbs moisture) require prebake of an hour at 125°C (single-layer structures; longer for multilayer or RF) to remove moisture, followed by almost immediate soldering—no more than a 15-minute delay from bakeout to solder exposure.
- Standoff—separation between hardware (connector body, for example) and the first FPW layer—may require a spacer to assure uniformity.
- Protrusion of contact pins beyond the solder fillet is frequently specified; a typical figure is 0.062 in. To meet this requirement the pins must be clipped at the correct length to provide standoff plus FPW plus fillet plus protrusion. This takes a shearing tool which must be kept sharp to minimize burring, because burrs interfere with assembly of FPW to hardware and may damage the insulation or pad.
- Conformal coat is neater if a simple mold or dam is used. This may be a silicone-rubber snap-on or other open-faced device.
- Connector pins should be predipped in the correct solder alloy to wash off gold, if present, verify solderability, and aid quick FPW attachment.

- Some level of preassembly inspection is wise; FPW and all hardware items should be known to be good before they're joined.
- Forethought to inspection and test may avoid excess labor. Thorough electrical test (shorts and opens) of finished assemblies will be performed; provision for probe contact is needed.

Before assembly work starts, responsibility for and ownership of scrap should be decided. If the customer provides the hardware, and an FPW assembly is rejected because there's a defect in a connector, who pays for the FPW and assembly labor?

Hand technique. Figure 11-1 is a typical hand-solder station. Arranged within easy reach around a long-working-distance, 10× microscope are isopropyl alcohol in a safety dispenser, a needle-tipped flux dispenser, two temperature-controlled irons with tip cleaning sponges, an easily cleaned glass working surface, a low-pressure air blowgun, coils of solder in solid-core and flux-core construction, wicking wire for removing solder, tweezers, and hand knives. The operator wears finger cots and a clean-room lab coat.

A hand-solder fixture is shown in Fig. 11-2. This is a soldering jig—it locates connectors and FPW in a desired relationship before

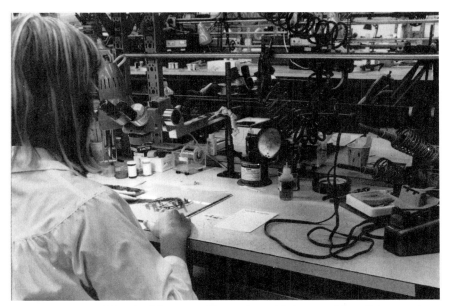

Figure 11-1 Typical hand-solder assembly station with long-working-distance microscope, soldering irons, flux dispenser, solders with several alloys, cleaning solvents, hand tools, and air line. (*Courtesy Teledyne Electronic Technologies.*)

Figure 11-2 Solder fixture frame. (*Courtesy Teledyne Electronic Technologies.*)

they're soldered together to assure conformance with assembly drawing requirements. Each of the 16 short FPW circuits ends in a stiffener/plated throughhole (PTH) pattern which pins onto the edge of the fixture; the large connector drops into the recess in the rear. An over-center clamp secures the main body in the fixture. The short circuits are formed in the small jig in Fig. 11-3 to provide a degree of adjustability in the location of each head pattern compared with the main body—without sharp crimps, short pieces of FPW can be very rigid. Figure 11-4 shows another use of crimping to improve formability.

Jigs of this sort also protect FPW and joints from movement and damage while the molten solder is applied and cools.

Mass methods. Mass joining techniques lower cost and improve quality by controlling process parameters and reducing the required skill. Tightened temperature control in conveyerized reflow, fixtured wave, or laser soldering allows use of thermally sensitive dielectrics such as polyesters (see Chap. 5) if contact with hot solder is restricted to the immediate joint area by fixturing. When 0.005-in or thicker polyester is used, with automated exposure times of 5 seconds or less and rational quality requirements which allow slight evidence of dielectric melting, high-volume, low-cost assembly can be enjoyed.

Assembly 229

Figure 11-3 Small forming jig. (*Courtesy Teledyne Electronic Technologies.*)

Figure 11-4 Detail of 93-Ω stripline flexible circuit cable. Periodic quilting provides controlled distortion to allow severe flexing without discontinuity.

Heating methods. Central to all mass soldering techniques is the method for controlling and applying the heat energy. The earliest technique was hand dipping into a pot of molten solder; this was soon improved by the addition of "boats"—enclosures which protected the circuitry from exposure except at termination points. In the mid-'60s, the hot soldering process was automated further by introduction of commercial wave soldering machines. Conveyer systems and further development of masking fixtures followed.

Solder preforms, automated torches, and carousels appeared, then the beginnings of surface-mount technology with hot-plate and IR solder reflow, followed by vapor-phase and finally hot-air soldering. Today, the wave solder technique appears to be receding while hot-air (convection) ovens take over.

Infrared is a powerful, rapid, clean heating method which, like laser soldering, suffers from optical problems and control issues. IR ovens can be used for reflow, but control of temperature is strictly guesswork. The procedure is to test conveyer speeds and power levels with dummy assemblies until the desired temperature profile is reached, then maintain identical conditions including color, roughness, masses, and so forth. Because the source—the IR emitter—is far hotter than the desired assembly temperature, small variations in the appearance of the FPW or RF can result in rapid and disastrous overheating.

Polyimide films absorb IR with great efficiency. This factor combined with very low thermal mass means that FPW areas must be masked if IR is used for heating.

Forced hot air provides best control and consistency. This process can't exceed set temperatures regardless of change in the assembly that's being soldered. It's slower, but adequately fast, it is safe, and it is becoming the popular general-purpose technique.

Surface-mount assembly on FPW

As the size of electronic components shrinks, mechanical and thermal issues in printed wiring design recede in importance. When vacuum tubes and high-voltage power supplies were leading-edge technology, structure was important, thus the wide use of glass-reinforced printed wiring boards (PWBs). Starting perhaps 15 years ago and constantly expanding into new equipment areas, the surface-mount technique is displacing throughhole attachment. It's attractive because it allows higher packing density and lowers assembly cost with high levels of automation. Surface-mount assembly (SMA) components are smaller, and other package elements—including FPW and PWBs—are also smaller. Heat rejection is lower with less power consumption, thus temperatures are reduced, and overall weight is sharply reduced as well.

The idea of directly mounting components on FPW seemed crazy at first because the primary characteristic of FPW is flexibility; it's never thought to have structural strength. But strength is relative: as component size shrinks, the need for support shrinks even faster, and today direct SMA on FPW is common.

Figure 11-5 shows a small, automated component loading device applying chip capacitors and resistors to an FPW circuit. The circuit is nested into a carrying pallet which supports and locates it accurately under the head. The pallet is a two-layer assembly of routed PWB laminates. The top layer has windows which accurately fit the outlines of the circuit, and the bottom layer supports the circuits and pins to the component loader X-Y table.

Each circuit, which is stiffened in the two connector areas as well as under the component sites, is fully outlined. A "chip shooter," which has capacity of 1 component per second, is fed from either tape bandolier magazines (at the top of the picture) or by vibratory feeders which are out of sight at the bottom. Components are dropped into prescreened solder paste but aren't otherwise bonded prior to solder attachment.

Figure 11-5 Chip shooter loading surface-mount components on FPW.

Figure 11-6 Conveyerized convection oven with a pallet of FPW.

Figure 11-6 shows the pallet emerging from a convection oven. The heat cycle takes 7 minutes at an air temperature of 450°F. Belt movement is very gentle to reduce the chances of disturbing a jointor jolting a circuit out of its pocket in the pallet.

Pressure connection

There are many commercially significant solderless assembly methods for FPW. These include individual and mass crimp as well as direct pressure in the form of low insertion force (LIF) and zero insertion force (ZIF) connectors.

Crimp connection is a high-volume, low-cost method for joining hardware to FPW conductors. Good crimp contact design provides stored energy to preserve high normal forces throughout a range of thermal stresses and over a period of time. Supplied in bandolier form and applied by high-speed equipment, crimp connections are low-cost and reliable, if somewhat bulky, and well suited for commercial use. Figure 11-7 shows a series of crimped contacts at the end of an FPW circuit; the rolled-over engagement tangs store energy for long-term reliability.

LIF and ZIF connectors are the rule in attachment of keyboard and switch array FPW circuits made by the polymer thick film (PTF) process. These connectors open up to accept the FPW, then cam shut

Figure 11-7 Crimp contacts on FPW.

to apply high contact force directly on the conductor runs. Insertion forces and wear are very low, which means these are the correct connectors for PTF circuit use. When a direct pressure connection is made to etched FPW conductors, they must have a protective, oxide-resistant surface finish for contact stability. Tin-lead is frequently chosen and is easily added during the manufacturing process or afterwards by hot-air surface leveling (HASL). Pressure connection via screws is seldom used and, when it is, requires torque isolation to protect the FPW from damage. A washer with torque reaction tail or captive contact is suitable.

FPW with spring metal conductors—beryllium copper is a good choice—can be formed to create a one-piece conductor/contact structure. The contact design may be conventional cantilevered arms or internal blades in a ring. An example of cantilever design is an FPW circuit which is formed into a tight U or hairpin shape, then forced onto the edge of a PWB where it functions as an edge-card connector, contacting both surfaces. Figure 11-8 shows such a circuit—a short FPW with beryllium copper conductors and fully bared contact areas at each end, in formed and flat condition, with the forming tool.

There are connectors which directly interface FPW with PWBs or wire-wrap pins. An example is a stainless-steel sheet metal cage with cantilevered spring fingers which is crimped onto a PWB, forming a

Figure 11-8 Forming fixture with unformed and formed U circuits in beryllium copper.

socket to receive the FPW. The fingers bear against the insulated base side of the FPW, urging exposed conductor surfaces on the opposite side into tight contact with PWB pads. A special tool is inserted into the cage to hold the fingers up while the FPW is inserted. The cage aligns FPW and PWB pads and locks into perforations in the FPW dielectric to strain-relieve the junction.

Bolted-together clamping plates which force plated-up contact areas on an FPW circuit against a matching pattern of contacts on a PWB or second FPW have been designed and used in specialized applications. Compliance and accommodation for nonplanarity, particularly with long-term thermal aging, are concerns with these designs as is the considerable mass and cumbersomeness of the clamp structure.

Similar technology is employed in other embodiments of direct pressure connectors which align and lock FPW to other FPW or PWB terminations by means of molded plastic housings and steel springs. In some designs the PWB is supported by the housing; others rely on the stiffness and flatness of the PWB for uniform contact force. In these latter designs, if the PWB bends or distorts as a result of long-term high-temperature use, contact force may drop below the minimum required for reliable contact.

A patented DCC-Post* direct-pressure design consists of two molded plastic members and a beryllium copper finger spring which lock onto an FPW by means of alignment pins, forming sockets to receive wire-wrap posts or male contact pins. FPW conductors are aligned to

---

*Trademark of Miraco Inc.

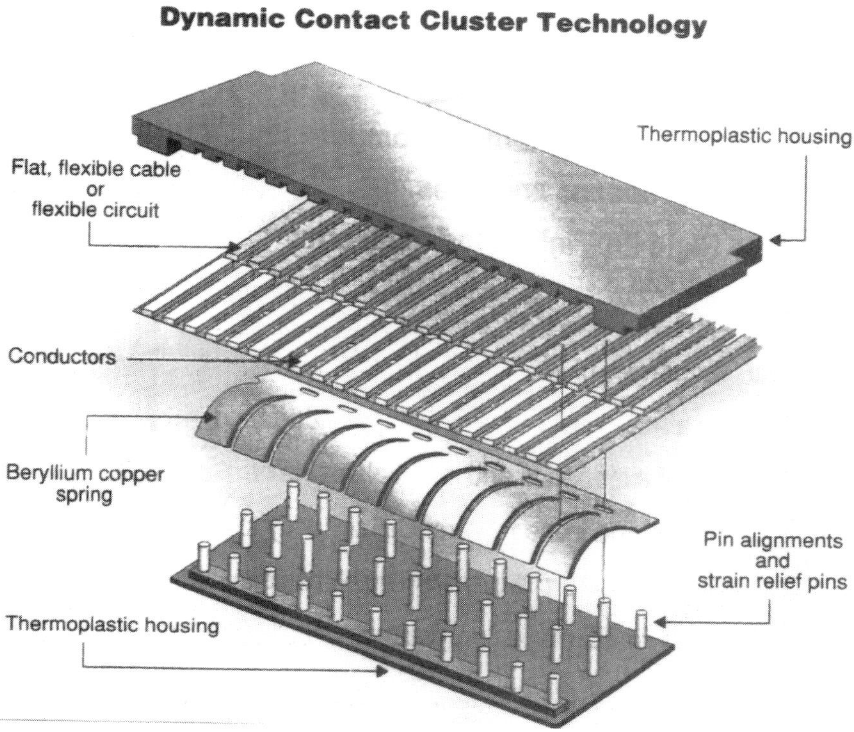

Dynamic Contact Cluster Technology

Figure 11-9 Schematic illustration of DCC-Post connector for FPW or tape cable, showing alignment pins, beryllium contact springs, and plastic housing. (*Courtesy Miraco Inc.*)

the posts and urged into firm contact by the fingers. This design isn't affected by the stiffness of the PWB dielectric. See Fig. 11-9.

Other methods

Thermocompression can be used to form permanent attachment between FPWs or between FPW and connectors. This is the most rugged and durable interconnection—essentially a weld—and is formed by automatic equipment at minimal cost.

Thermocompression temperatures for copper are quite high. Insulation charring will occur, even with heat shielding and pre-stripping. The process can be used on fully insulated FPW if a correct insulation choice is made. Thermoplastics such as polyester and FEP Teflon* work well with thermocompression; they displace neat-

---

*Registered trademark of DuPont.

ly from the bond site with minimum char. Thermosetting systems aren't sufficiently displaceable and require prebaring of the bond site.

The process results in conductor forging to roughly 50% of original thickness, thus creating a weakened section. Since all the insulation is displaced from the bond area, further weakening the area, potting or other reinsulating/reinforcement is required.

Conductive adhesives (see Chaps. 6 and 8) are rapidly gaining acceptance as alternatives to solder or pressure connections. When they're used, absence of high assembly temperatures allows choice of low-cost FPW dielectrics such as polyethylene terapthalate (PET) and polyethylene napthanate (PEN) for synergistic cost benefit.

It's claimed that conductive adhesives are more reliable than solder because they are more elastic, therefore survive thermal cycling better. It's certainly true that assembly is less expensive, and for applications which do not require low resistance or significant current-carrying capacity, conductive adhesives hold considerable promise.

Forming

Tools

FPW forming may be done before or after solder or other operations. Simple jigs and fixtures are sufficient to control the location and shape of bends. Figures 11-10 and 11-11 illustrate two approaches. Figure 11-10 shows a forming fixture which locates the connector on tooling pins, grips the connector with a stripper-like plate (springs not visible), and forms the FPW downward at the desired dimension over a smoothly radiused edge, using a semisoft plastic pusher to minimize insulation damage.

Figure 11-11 shows a similar fixture; this one crimps the FPW to provide length adjustability. The swinging arm with forming pins is pressed downward onto the FPW, driving it in between a second set of pins (not visible below the FPW).

Typical features in both these fixtures are

- Connector (or FPW termination pattern) is pinned and clamped to establish an accurate dimensional reference.
- Forming surfaces are smoothly radiused and/or soft to minimize insulation damage.
- The FPW is free to slide and accommodate itself to the forming action. (Never clamp both ends, then try to form the middle.)

Assembly 237

Figure 11-10 Large forming jig.

Figure 11-11 Small forming jig.

Heat, permanence

FPW dielectrics are mixtures of thermosets and thermoplastics. Thermoplastic systems can be heat-formed with some precision, because these dielectrics reform at modest temperatures to produce a stress-free shape. Dielectrics which are somewhat thermosetting—polyimide films with their adhesives—do not form well and will attempt, slowly, to revert to their flat, unformed condition.

It can be dangerous to heat-form polyimide systems: buckling, undesirable wrinkles, and adhesive movement could result. An exception is the application of very firm pressure and elevated temperatures to small areas specifically to force adhesive movement. When this is done correctly, enough adhesive can be displaced—but only a short distance—to create a permanent and tight bend. Springback of roughly 15% is normal.

It's not a good idea to assemble polyimide-insulated FPW with tight, clamped bends if elevated temperatures (80°C or above) or long dwells are expected. Wrinkles or delamination will result from the combination of standing stresses and heat.

Forming note: Because FPW always springs back to some degree, it's not possible (or reasonable) to form FPW precisely. If a formed shape is shown on design documents, be sure the notation "shown in restrained condition" also appears.

Designs which have the maximum metal at the forming site spring back least. When a tight bend is needed, all possible conductor metal should be left in the circuit—even in a second dummy layer, if possible. Figure 11-8 shows a three-piece fixture which forms short FPW strips into a U shape to create a top-to-bottom interconnector for PWBs. In this circuit, the cross section is 50% conductors, which helps to hold the formed shape.

Potting

Potting is an art that should not be approached casually. Accurate mixing and dispensing of the compounds, creation of neat, smooth molded contours, absence of bubbles or voids, high insulation resistance, and good yield don't happen without skill.

There's an endless variety of potting mold designs and procedures. Here are guidelines:

- Potting workstations should include a hot plate, ultrasonic cleaner, hand tools (spatula, knives, tape, scissors, tweezers), scale or balance suitable for the size of the pours (or freezer, for prepackaged compounds), vacuum-degassing equipment, and drying/conditioning and curing ovens.

- Connectors must be sealed before they're potted to prevent penetration of uncured compound into the contacts. Connectors have one half which has floating pins that align with the other half. If the floater is soldered to the FPW, trouble abounds: floating contacts aren't sealed and assembly to FPW will stiffen the contacts, defeating self-alignment. The best technique is to solder the fixed half (usually the male) onto the FPW—this avoids both problems. But connector choice is usually determined by what's on the other side of the interface (see Chap. 6 section, "Hardware"). If assembly must be made to the floating half, then (1) the contacts must be sealed to prevent potting compound from flowing through and fouling the mating surfaces, and (2) the two halves should be mated during the sealing and assembly process to centralize and align the contacts.

Molds

Designing potting molds is a complicated task. A typical mold is built up of several parts which are disassembled to remove the finished FPW assembly. Mold design follows these guidelines:

1. The connector seats firmly (with gaskets) into the mold and is secured by temporary screws.
2. If FPW enters through a sidewall, it's gasketed and clamped as well.
3. Walls of the mold defining the internal volume are smooth, with draft to ease demolding.
4. An entry port and an exit port are needed. Potting compound is injected through the entry port, flows through and around the FPW and hardware to flush all air out, then exits through the exit (vent) port.
5. Gentle heating to lower viscosity and assist in air expulsion is commonly used. Aluminum is a good mold material choice for easy machining, handling, and good thermal conductivity. However, if long life is expected, aluminum isn't a good choice because it's difficult to clean without marring.
6. Molds must be disassembled to allow the connector and FPW to be extracted and to allow polishing and cleaning if necessary. Mold release is reapplied as needed.
7. It's a good idea to include identification marks.

Materials

Potting compounds are somewhat plasticized to reduce brittleness and cracking which could result from differential shrinkage during cure. Softer compounds are more gentle on FPW and hardware terminations, less likely to cut or damage FPW insulation during severe bending or flexure. Polysulfides are popular with airframe manufacturers because they're fuel-proof; polyurethanes and nylon-flexibilized epoxies are widely used as well. Most compounds are opaque; cure of some types is inhibited by traces of foreign chemistry or mold release. The tendency to form voids or bubbles varies with type (and application temperature).

Vacuum degassing is used with complicated hardware and potting shapes that don't readily fill. Certain potting compounds are more likely to retain bubbles or to react with contaminants (sometimes the mold release) during cure. The vacuum fill process consists of placing the assembled FPW, hardware, and mold, and the mixed compound in a vacuum chamber and pumping it down. After the compound has ceased to outgas, it's poured into the mold by remote manipulators. Spreading the compound out into a thin sheet en route to the mold helps to release any trapped volatiles; this is accomplished by the use of specially shaped funnels.

Injection mold process

Injection molding is used in high-production designs. This technique requires a good deal of mold development and an FPW-hardware unit that can stand the high temperatures and pressures required. When well executed, injection-molded assemblies are neat, efficient, and long-lasting.

Major engineering issues are

- Sealing the mold onto the FPW
- Correct placement of inlet and vent gates
- Determining correct injection temperature and pressure
- Choice of molding compound—preferably a low-temperature-cure material with excellent electrical properties

Conformal Coating

Solder joints and cutoff connector pins have sharp edges or points which could damage FPW insulation or, at the least, catch foreign material. Conformal coating, a sort of buttering-on of insulation, is used to cover and protect arrays of solder joints where potting isn't needed. See Fig. 11-12.

A highly thixotropic compound is preferred, that is, a material which can be easily pushed or smoothened into a desired shape but doesn't slump or flow afterwards. A totally bubble-free enclosure isn't

Assembly 241

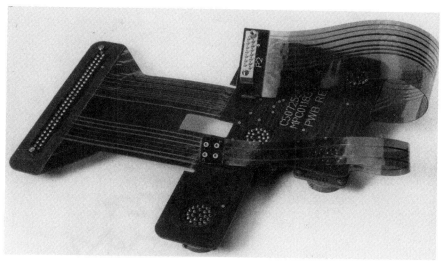

Figure 11-12 Military rigid-flex assembly uses a conformal coating to insulate the solder joints. (*Courtesy Parlex Corp.*)

expected with a conformal coat—the idea is to cover the joints, leaving a smooth outer surface. Good choices include non-acetic-acid-curing silicone rubber [so-called room-temperature vulcanizing (RTV) material] or thixotropically loaded epoxies or urethanes.

Nomenclature

Built-in part numbers, revision levels, and identification of connectors and terminals is one of the inherent advantages of FPW. In cases where it's impossible to include such markings in the artwork, surface stamping or marking is used. Application of nomenclature requires adequate preparation of the FPW surface by solvent cleaning, mixing of ink (reactive inks for military FPW), setup of the appropriate rubber stamp or screen stencil, and application in the correct area followed by ink cure.

Nomenclature can be applied to labels, which are then transferred to the FPW assembly. Bar coding is a form of label nomenclature application.

Several photographs (Figs. 4-14, 4-20, 8-26, 8-33, and 9-11) show external nomenclature. MIL specifications dictate "nonconductive, permanent, fungistatic ink or paint" which is applied to the circuit or to a label which is applied to the circuit. Note that these specifications require that external markings must resist flux, molten solder, cleaning solutions, and conformal coating.

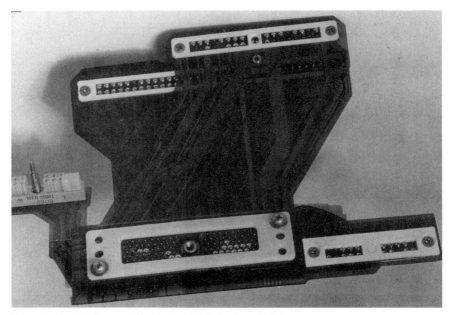

Figure 11-13 FPW assembly with molded strain-relief rings.

Mechanical

There are occasions where mechanical assembly work is performed on FPW. Attachment of stiffeners or connectors may require eyeletting or bolting. The FPW itself may be adhered or clamped to the walls of electronic housings to keep it clear of cooling fans or moving devices or for enhanced shock and vibration resistance. Dynamic applications frequently include careful positioning and clamping of FPW to assure long life. Multiple unbonded layers of FPW may be joined together for neatness, or may be bundled together with loose, interleaved shields for improved performance. An example of mechanical assembly is shown in Fig. 11-13, a view of the solder side of an FPW assembly. The white rings are structural plastic moldings engaged by the connector mounting screws to provide support and to clamp together the multiple layers of FPW. The moldings also act as strain reliefs, protecting the solder joints from flexure.

Application of conductive taping as an external shielding technique is a mechanical process. Taping is also used to bundle oversized wire conductors or coaxial cable together with FPW to satisfy small-lot requirements or engineering change orders.

Summary

The method of assembly has impact on FPW design and circuit cost as well as on assembly cost. Solder attachment, the most common technique, forces use of higher-cost dielectrics which can withstand the necessary temperatures. An automated solder process, with adequate shielding and reasonable inspection requirements which allow some evidence of melting, may allow use of low-cost PET, vinyl, or other dielectric for reduced assembly cost.

The SMA technique is increasingly used; smaller components and more sophisticated equipment design facilitate the trend. Pressure contact via custom connectors or conductive adhesive where adequate are other methods which allow use of low-cost dielectrics.

Forming is possible, particularly in thermoplastic dielectric systems. Polyimides can be formed to some degree but tend to creep and distort with time. No formed FPW shape can be produced to high precision.

Potting and conformal coat (see also Chap. 5, "Dielectric Materials") are used to reinsulate terminated areas or to protect adjacent FPW from contact with sharp solder joints or clipped contact pins.

Nomenclature in the form of printed or stamped markings is applied as part of assembly process to aid in identification of connectors and contacts and for field repair.

Chapter

12

Examples of High-Volume and Unusual Flexible Printed Wiring

Military Reel Cable

The single-layer flexible printed wiring (FPW) in Fig. 12-1 is 50 ft long. In use, it's stored on a reel and unwind/rewinds to follow the motion of a navy optical device.

The dielectric is 0.005-in polyester with 0.002-in polyethylene "adhesive," chosen because it's compatible with the fusion process, has a smooth, slippery surface for reduced friction on the reel, and is mechanically tough to withstand shipboard life. Conductors are 0.0028-in rolled-annealed (RA) copper foil.

The design consists of a series of parallel conductors with connector termination patterns at both ends. Except for the connector flare-outs (one is visible in the picture), this application would be well-served by flat conductor cable. Custom ends and stringent MIL specifications precluded anything but fusion-bonded, one-piece FPW.

Imaging was performed by screen printing in roughly 8-ft lengths of conductor pattern, skipping an 8-ft section, then screening again, skipping again, and so forth to the ends, where the custom patterns were applied. Once the interrupted imprint was dry, missing sections were applied by a second printing process.

Etching and image stripping were straightforward using conveyerized equipment. Coverlayer lamination required multiple overlapping press cycles which, because this system bonds by the fusion process, smoothened into each other. Inspection and test included "wet sponge" short-circuit-to-ground wiping over the outer surfaces to verify a saltwaterproof dielectric.

Figure 12-1 Military 50-ft-long reel cable. Insulation is polyester-polyethylene composite. Cable was produced by print-and-etch technique with skip-screen printing. (*Courtesy V. F. Dahlgren.*)

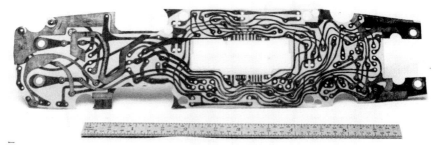

Figure 12-2 Early telephone handset circuit. The dielectric is nonwoven fiber mat with flexibilized epoxy. The circuit is double-sided with plated throughholes; throughholes and outline were punched. The surface finish is nonreflowed tin-lead. (*Courtesy V. F. Dahlgren.*)

High-Volume Epoxy-Mat FPW

A two-sided, epoxy-mat-based, PTH-interconnected circuit (Fig. 12-2) was manufactured in high volume by the panel process for telecommunications use. The mat-reinforced dielectric and plasticized resin allowed die-punched plated throughholes (PTH). Etch resist and surface finish on the etched runs is unfused tin-lead. Composite layout consisted of individual circuits in spaced rows. Finished panels were cut into strips, then aligned with and fed through a die which cut both the outline and nonplated throughholes.

High-Volume Vinyl FPW

Another high-volume circuit for telecommunications was made in fused vinyl/copper foil laminate (Fig. 12-3). Base lamination, screen image printing, and etching were performed by the roll-to-roll technique; image stripping, coverlayer alignment, surface preparation (oxide treatment), and lamination were in panel form. Coverlayer apertures are very large to compensate for postetch material movement and coverlayer flow; the surface finish is electroless tin.

Stripline

This complex product was the cover feature for an industry magazine. Overall length of the 93-Ω stripline FPW circuit was 10 ft (Fig. 12-4). Each end was hand-soldered to a paddle board connector—a small printed circuit board (PCB) with edge contacts.

The conductor layer contains eight 0.008-in-thick×0.025-in-wide conductors on 0.1-in centers, oxide-treated and roll-laminated between polyethylene base and coverlayer films in continuous lengths of 100 ft or more. Shields were panel-processed on polyethylene foil laminate and coverlayered with polyethylene.

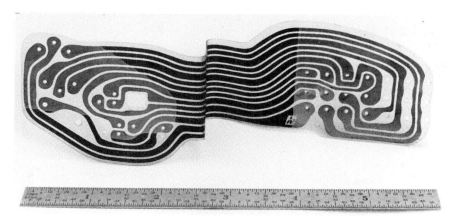

Figure 12-3 An early roll-to-roll process produced circuitry for telephone timing equipment. It was manufactured in high volume on fusion vinyl/copper laminate with a press-laminated vinyl covercoat and tin-lead conductor finish. Note application of the area baring concept. (*Courtesy V. F. Dahlgren.*)

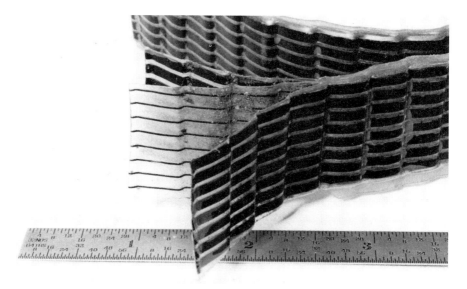

Figure 12-4 Detail of a 93-Ω stripline flexible circuit cable.

Lamination was performed in long, narrow aluminum plate fixtures in a custom 11-ft press. Sharpened needles in the base of the fixture pierced between shield conductors to align the layers on the foam dielectrics.

Conductors were sized to meet conductivity requirements. The resulting width forced use of foam with a dielectric constant near 1 to yield 93-Ω impedance in a reasonable thickness. Fusion bonding joined the three circuit layers and two foam strips in a brief exposure to modest temperatures below 300°F.

Nominal shield-to-shield separation required for 93 Ω, even with the foam dielectric, was about 0.15 in. A three-layer FPW can't be bent around radiuses less than 25 times the overall thickness. Because of the low mechanical strength of the foam, plus the effect of uncontrolled and varying shield separation on impedance which would result from bending the circuit, a novel technique for stabilizing impedance and improving flexibility was used. As shown in Fig. 12-4, the shields are periodically crimped, or quilted into the foam, to produce a spaced series of areas where thickness is reduced and flexibility increases. When the cable is flexed, bending occurs in these prebent areas, which are spaced closely, compared with signal wavelength, to minimize signal degradation.

Examples of High-Volume and Unusual FPW   249

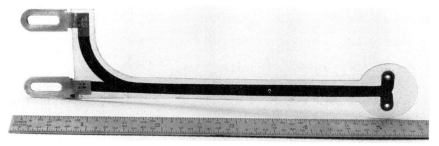

Figure 12-5 High-voltage ignitor cable. (*Courtesy V. F. Dahlgren.*)

High-Voltage Ignitor Cable

Fusion-bonded, two-layer circuits insulated with Kel-F (an early fluorocarbon polymer) were made for a military program (Fig. 12-5). Soldered eyelets at the round head end and spot-welded ears at the opposite end provide easy and reliable interconnections. Built in considerable numbers and several designs, these high-voltage cables were 100% acceptance-tested at 5000 V dc.

Kel-F* in its normal condition resembled Teflon, and formed the basis of much of the early fusion-bonded FPW. However, if it was allowed to cool slowly from bonding temperatures—450 to 500°F—it developed a crystalline structure, losing its pale bluish color and tending to form microcracks. The technique for avoiding this danger was to shock-cool after fusion bond by extracting a platen-load of circuitry from the hot press and dropping it directly into a pan of cold water!

Vacuumtight Stripline

In a 48-in-long 50-Ω stripline FPW cable, solid dielectric and midbody potted into a flange connector create a vacuum tight feedthrough for use in vacuum instrumentation (Fig. 12-6). The circuit contains 60 conductors which terminate via zero insertion force (ZIF) connectors at each end.

Shields interconnect with each other and ground conductors in the conductor layer by means of via holes filled with polymer thick film (PTF) in closely spaced rows along each outer edge and down the center.

---

*Registered trademark of DuPont.

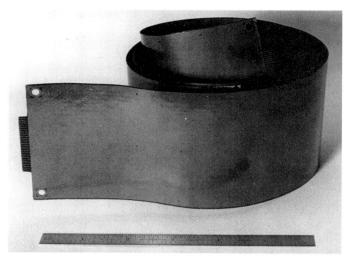

Figure 12-6 Vacuumtight stripline FPW.

Outsized Backplane

A 4.4-m-long (14.4-ft), 0.44-mm-wide (1.75-in) 18-layer backplane was built in medium production volumes for use in a scientific particle detector. Spaced accurately—to 0.2 mm tolerance—along the top surface are 56 termination patterns for attachment of semiconductor-based particle detectors (Fig. 12-7). The detectors are positioned and soldered to the patterns with less than a millimeter of error by use of giant coordinatograph. Each termination pattern is long enough to allow for slight repositioning.

Also visible in Fig. 12-7 are tooling pins [a pair located every 20 cm (8 in)] used to align outline tools and to register and secure the layers for lamination. Each etched layer contains postetch punch (PEP) targets which are punched to create tooling holes to engage these pins. Rough-trimmed for processing, layers were 6 in wide; the finished backplane is 44 mm (1.75 in) wide and 4 mm (0.157 in) thick.

The backplane contains four layers of power and test circuitry (layers 2, 3, 4, and 5; layer 1, the coverlayer, was deleted early in the program to lower cost). These layers are followed by seven pairs of signal and ground layers. Each backplane has 336 signal conductors at about 35-Ω impedance. There's no layer-to-layer interconnection; this is simply a well-registered, bonded stack of FPW layers with interposed shields.

The backplane is made of 0.005-in polyester film with 0.0014-in RA foil conductors, bonded together by polyester adhesive film layers.

Figure 12-7 Close-up of backplane termination pattern with tooling pins.

Polyester construction nicely matched both product requirements and manufacturing capabilities:

- It's inexpensive. Each backplane requires almost 150 ft^2 of laminate and adhesive. Since these backplanes are used in groups of hundreds, material cost is very important.
- Overall maximum thickness of 0.157-in for 18 layers allows 0.008 in per layer, which is satisfied by 0.005-in polyester base film, 0.0014-in copper foils, and 0.001-in adhesive layers, assuming slight compression into the etched patterns.
- Signal layers are heavily shielded and exceptionally long (over 10 ft), therefore low dielectric constant and low losses are necessary for good signal transfer from the low-level semiconductor detectors to output connectors. The design was intended to present 50-Ω conductor impedance, but actually resulted in 35 Ω. Of the two techniques for increasing impedance (narrowing the runs or increasing the separation between shield and run), neither can be used here—conductor width is fixed by resistivity requirements and no space is available for adding insulation because the backplanes are at maximum thickness as built. Fortunately, the semiconductor drivers were also not to specification. They are 25-Ω devices, thus adequate signal transmission was realized.
- As hitchfeed lamination is required, a low-temperature thermoplastic bonding dielectric is preferred.

- Extremely long conductor runs at close spacing (0.015 in) require excellent surface resistivity. Polyester adhesives are very good insulators.
- Since soldering is performed only at large connector output pads and onto even larger pads for detector attachment, the low thermal resistance of polyester is acceptable.
- Use environment is benign, with no temperature extremes, no vibration or shock, no flexing.

Bonding adhesive is tacked in place on each layer prior to die cutting to eliminate alignment concerns. Each layer is cut prior to lamination by steel rule dies (43 required to produce each backplane) which partially predefine the outline (as shown) and create crenelations in overlying layers to expose terminal pads in other layers. Each die is aligned with the termination patterns via PEP tooling holes.

Lamination is performed in a 16-ft custom laminating fixture by using two cycles of a 10-ft laminating press.

Not visible in Fig. 12-7 is the cutoff design, which needs some explanation:

Each backplane has termination patterns for 56 detectors

Each termination pattern has six signal conductors

Each 48-conductor signal layer terminates eight detectors

Signal layers are 6, 8, 10, 12, 14, 16, and 18

Shields are 7, 9, 11, 13, 15, 17, and 19

To reduce material consumption and cost, each signal and shield layer pair extend only to the end of its termination pattern. For example, layer 6 and its shield, layer 7, provide 48 signal connections to the first eight detector patterns nearest the connector end of the backplane; since no conductors continue beyond, both are deleted between patterns 8 and 9. Layer 8 (and its shield, layer 9) passes underneath to the next group of eight termination patterns and so forth. The four common layers—2 through 5—were present throughout, so that the farthest end of the backplane consists of six layers (2, 3, 4, 5, 18, and 19) for a combined thickness of somewhat more than a millimeter.

Accommodation for differing thicknesses down the length of the backplane and the extremely deep crenelations requires a stepped laminating pad construction plus 0.25 in of conforming pad material.

Vacuum storage for 24 hours prior to lamination is required to extract all traces of final cleaning chemistry and rinses which otherwise degrade conductor-to-conductor resistance (a particular problem in layer 18, in which there are over 13 ft of conductors spaced 0.01 in apart).

Figure 12-8 Bandsawing the backplane.

Figure 12-8 shows the bandsaw outlining procedure. The desired width is 44 mm (1.75 in) with 0.2 mm (0.008 in) tolerance. Conventional routing is not practical because of the extreme length and highly thermoplastic, router-fouling dielectric. Brief research into machining procedures determined that bandsawing is quick, clean, and adequately precise. The picture shows the precision sliding support, repositioned every 4 ft, which controls width.

The bandsaw tooth configuration was chosen to produce a slightly irregular surface. This roughness both improves adhesion and protects the silver conductive-ink edge painting which interconnects all shield planes uniformly along both edges.

Figure 12-9 shows a backplane stretched out for attachment of connectors which are lap-soldered to an unbonded area of each signal layer by the worker at the far end.

Handling these outsized circuits through a typical FPW factory is an interesting challenge. In laminate and layer form, an entire back-

Figure 12-9 Stretched-out backplane.

plane can be rolled up and carried between processes in a single tote tray. Once it is bonded into a semirigid, 15-ft circuit, transportation of each circuit between departments requires several workers, prearrangement, a preplanned route, and generous doors and hallways. Eventually it was learned that the circuit would tolerate some bending; minimum size of an outlined circuit was 5 ft in diameter.

High-Volume Miniature Backplanes

An early unbonded, multilayer backplane was built in 10 different designs in quantities of several hundred each. Each 6.5-×0.7-in "stick" contained up 12 layers with 300 eyelet interconnections 0.046 in in outside diameter.

The dielectric for the conductor layers was 0.005-in fusion FEP Teflon* in base and coverlayers. The base side of the 0.0014-in foil was oxide-treated but the coverlayer side was not. The coverlayer was a full sheet of FEP, with pads bared after coverlayer lamination by

---

*Registered trademark of DuPont.

Examples of High-Volume and Unusual FPW   255

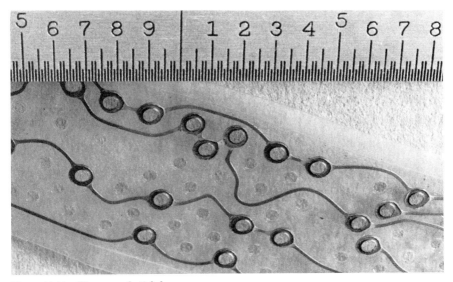

Figure 12-10 Close-up of stick layer.

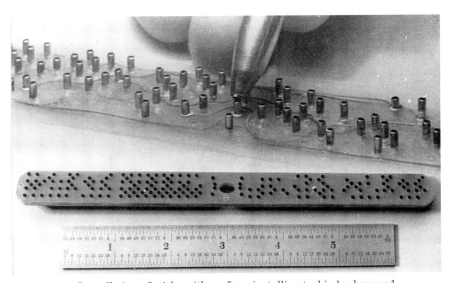

Figure 12-11 Overall view of sticks with preform installing tool in background.

careful hand cutting; absence of tight bond to the foil allowed easy removal of the cut-off coverlayer (Figure 12-10).

The layers were imaged and coverlayered in panels, then cut down to individual pieces for finishing. These operations included eyeballed punching of pad holes, master pattern punching of clearance holes, and 95-5 hand soldering to the eyelets using preforms (the background of Fig. 12-11 shows the preform installing tool). A projecting tab at one end provided space for inspection stamps and included a set of punch-out numbers for serialization.

The $\frac{1}{32}$-in epoxy-glass top and bottom boards were machined and drilled by numerical control (NC). Eyelets were copper, funnel-set in the finished board to lock the topboard in place, compressing all layers tightly together.

Up to 12 layers were used (limited by available eyelet length), with many double solder joint interconnections between layers.

Because of the unbonded construction and the large number of clearance holes and solder balls resulting from the attachment technique, foreign material trapped between layers constituted a major quality problem. It was addressed in brute-force fashion by visual inspection after each layer was soldered in place plus microscope examination of "wet" x-ray films of finished sticks.

Final electrical inspection was performed by a universal probe fixture using key inserts to pattern the active probes to the part under test. The controller for a "shorts and opens" test was relay-based.

Mechanical inspection was complicated by off-part coordinate origin. Extensive calculations were needed to determine if an origin could be found which made measured dimensions acceptable. In practice, since the top and bottom boards where NC-machined, all dimensions were well controlled; except for the hypothetical origin, mechanical inspection was straightforward.

Chapter 13

Summary

Purpose

The purpose of this book is to explain flexible printed wiring (FPW) technology in enough depth and detail to allow program managers and designers to develop successful designs and supervise procurement and assembly. This is not an engineering textbook and won't instruct in every nuance; it's a guidebook for managers.

Chapter 1: Introduction

FPW appeared as a commercial product in the late '50s. It was an unexpected outgrowth from attempts to improve printed wiring board (PWB) technology in military electronics. The search for a way to minimize cracking in adhesive-bonded printed wiring boards led to the use of the adhesive by itself as a substrate. The resulting printed wiring "board," although extremely weak and fragile, gave rise to the concept of a mass-produced universal interconnect device. This new product offered the technical and economic benefits of mass manufacturing techniques as used for rigid printed wiring, and, because it is flexible, could be routed around corners and throughout electronic packages as a replacement for wiring harnesses.

Early circuits were built by the fusion process using high-performance polymers in the fluorocarbon family. The difficulty of maintaining dimensions through this semimolten, high-pressure process led to the use of binary dielectric systems in which a higher-melting polymer provided structure and stability while a lower-melting layer bonded conductors and other insulating layers together.

Desire for shorter, lower-temperature bonding cycles and even better dimensional stability stimulated development of thermoset adhesive-bonded systems which became the mainstay of FPW production by the middle '60s.

Plated throughhole (PTH) technology, reduced to commercial practice in roughly the same time period, provided the means to build multilayer constructions including circuits with both rigid and flexible dielectrics—so called rigid-flex (RF) circuitry. These two developments—dielectric systems with adhesive-bonded films and PTH terminations—continue to the present when, worldwide, something like $1.3 billion in FPW and RF products is manufactured yearly.

FPW is a somewhat mysterious product to the uninitiated, with a large vocabulary of unique terms and jargon. As is typical of high-technology products, full understanding requires an appreciation for the intended meaning of these special terms.

Chapter 2: The Engineered Interconnection

Any electronic assembly needs structure and interconnections. Wire can provide interconnections; PWBs can provide interconnections and structure in two dimensions; FPW (with stiffeners) can do it all. FPW is the universal, fully engineered interconnector for all electronic applications. Much electronic equipment in use today—flying disk head interconnects, high-speed printers, low-cost keyboards—couldn't be built by any other method.

Compared with wire interconnects, FPW is much more complex, time-consuming, and expensive to engineer for the first piece, but rapidly pays back these costs in production use, where precise reproducibility, absence of installation or wiring error, compactness, weight saving, and easy service result in lowest possible cost. Wire is the efficient interconnect for low-volume applications or where the design isn't fixed. FPW is the mass production method; wire is the developmental method.

PWB offers many of these advantages, but, lacking ability to penetrate into the third dimension, is nowhere near as useful or versatile. FPW, in rigid-flex form, is the best of all interconnects, if costly to design and build. Here, in the ultimate production interconnect, is component support as provided by PWB, combined with full three-dimensional design and functional flexibility of FPW, available through cost-efficient high-volume manufacturing.

FPW saves weight and space because the conductor sizes are determined by electrical requirements, not convention or termination requirements. Thinner, smaller conductors require less insulation for further savings.

The planar structure of FPW maximizes flexibility in one plane but reduces it in others. Applications such as disk drives and printers take advantage of the low torque requirement and long life benefits of FPW through designs which exploit the flexible direction.

The ability of FPW to be formed, creased, and folded provides easy installation, even in cramped packages. Low mass, high strength, and good thermal resistance make FPW robust in military environments or anywhere that shock, vibration, and temperature extremes are experienced. Planar layout, consistent conductor patterns, and built-in terminal identifications make FPW-interconnected equipment easy to troubleshoot. Repair is difficult but possible.

Chapter 3: Manufacture of Flexible Printed Wiring and Printed Wiring Boards

Manufacturing equipment and methods for FPW are nearly identical with PWB practice save for crucially important differences:

1. FPW materials, because flexible, are less dimensionally stable. Where PWB materials include glass-fiber reinforcement, FPW dielectrics are based on polymer films. Reduced stability forces smaller panel sizes and more complex tooling alignment schemes, raising labor cost per circuit.
2. In-process sheets or panels of FPW circuitry are difficult to handle and process without damage, because they're very limp. Conveyerized equipment as used in PWB manufacture, for example, can be used but only with leaders.
3. FPW is almost always transparent and thus easily inspected. It is frequently rejected for embedded, unidentifiable "foreign material" which wouldn't be detected in PWB products.
4. A significant product difference between FPW and PWB is the almost universal use of coverlayering—topside insulation films bonded onto the conductor patterns—in FPW.

Coverlayers are functionally similar to PWB solder masks. The typical coverlayer is a high-performance polymer film and adhesive identical in composition and performance to the base dielectric layer.

Production of FPW is significantly more complicated, with lower yield and less certain delivery because of these differences. Familiarity with PWB technology could lead to a dangerous assumption of competence to manage FPW projects, an undesirable situation which can be avoided through study and appreciation of these differences.

Dimensional instability in flexible laminates forces use of small panel sizes and multiple tooling pins to reduce alignment variations to workable levels. In spite of decades of process development, FPW is very much a custom product. Per-piece prices are higher, practical tolerances wider, and subjective visual inspections more significant (because of transparency) in FPW procurement as compared to PWB procurement.

PWB quotes are available within hours of inquiry, and multiple sources will all price within a small range. In FPW procurement, delay of up to 2 weeks with technical discussions to clarify design specifics, choice of materials, and specifications is typical. Quoted prices from multiple vendors will range more widely, with greater variation in delivery, compared to PWB. Whether this is because the dimensional uncertainty of the laminates affects the confidence of company managers, or the processes are not as developed because so many different ones are used, or the industry simply has never settled into a volume production mindset is unknown, but the fact is that FPW production isn't as mature as PWB production. Designing and procuring circuits is a harder task in FPW than in PWB.

Chapter 4: Design

FPW design is a challenging, interactive process combining mechanical and electrical performance with producibility considerations. It begins with accumulation of a complete set of mechanical and electrical requirements together with a detailed picture of the intended assembly procedure, field-service requirements, and a cost target.

Mechanical information is used to create a precise mock-up onto which prototype "paper-doll" FPW can be fitted to explore possible shapes and layouts.

Electrical requirements are reorganized to produce listings of conductors by circuit type and by geographic similarity. Heavy current runs, which require thicker conductor foils, or sensitive runs, which need shielding, are grouped together for design consideration, as are all runs which interconnect specific area to specific area, so that an overall pattern of the most efficient use of FPW area can be seen.

As the cost of FPW is directly related to circuit area (specifically the number of circuits which can be produced on a standard panel), efficient design requires high conductor density in each FPW segment to minimize cost per conductor.

The method of attachment between FPW and hardware is a design factor which affects material choice, conductor density, and terminal design. The most common technique is soldering, which dictates high-quality dielectrics and adhesives which can withstand the associated thermal and mechanical stresses. If specialized joining methods—mechanical, conductive adhesive—are to be used, the conceptual design process remains largely unchanged but lower-cost materials can be used.

Termination areas are the densest part of a FPW design and therefore determine the maximum number of conductors which can be incorporated into a circuit. Once the overall pattern of interconnection is determined from study of similar circuit and geographic groupings, the next design step is to examine the dense areas—connector patterns or the like—to see how many conductors can be routed through the pattern to interconnect according to the wiring list.

Many routing techniques are used to reverse or interpose conductor sequences so that the pattern at one end of a bundle of conductors mates with the correct terminations at the opposite end. An important advantage in FPW design is that a conductor is not limited to two ends, as is the case with wire; FPW runs with multiple branches can interconnect multiple circuit points with the least number of joints. However, attempting to fit all these contact points into a single run may result in a cumbersome, low-conductor-density, high-circuit-area design that is cost-inefficient. Breaking up complex circuits into multiple pieces is unelegant but lowers cost. Where allowed, multiple solder joints on a hardware pin can be used to rejoin segmented runs; another method is the PTH process.

The PTH process increases connectivity dramatically, and when it is used, PTH can be used anywhere in the panel design without further cost. PTH is a powerful technique for compressing and condensing conductor patterns and, in rigid-flex design, allows use of rigid surface layers whose excellent thermal resistance serves as armor for more fragile FPW layers which are sandwiched beneath. High-density RF design usually requires high-performance FPW laminate materials from emerging "adhesiveless" technology.

Multilayer design

High interconnect density requires multilayer circuits. Design of these structures requires attention to alignment and registration issues. Typical methods for enhancing layer-to-layer alignment:

1. Artwork factoring to compensate for after-etch shrinkage
2. Multiple tooling holes and pins
3. After-etch optical punching of tooling holes [postetch punch (PEP)]
4. Small panels—individual circuits are the ultimate
5. Use of adhesiveless laminates, which have improved dimensional stability
6. Maximum use of stabilizing copper—wide borders, filled-in areas in the circuit

Rigid-flex design

RF design is challenging, but only in the amount of detail to be coordinated. Compared to an FPW multilayer, RF adds:

- Another outline—the rigid area has its own outline
- Multiple material uses—rigid outer layers are standard PWB materials; bonding adhesives may be PWB prepregs
- Added tooling for slot routs, windowing, and filler preparation
- Possibility of progression or bookbinder design

Tooling

Design includes tooling—drill programs and die designs for perforating and outlining, potting molds and forming jigs for finishing operations, laminating fixtures, and plating racks. It also includes choice of dielectric material and adhesive.

FPW is very cost-effective as a replacement for shielded and controlled-impedance wires. A variety of shielding techniques can be used with a range of added costs and effectivenesses. Controlled impedance is readily incorporated using established design techniques.

The most cost-effective strategy in cases of uncertainty is to test the simplest design. Because FPW is mass-produced, successful prototype designs can be duplicated for production use; if the test circuit passes requirements, so too will production circuits. Replicability validates use of empirical engineering.

Complete documentation of FPW design includes computer-aided design data for artwork plotting together with dimensional drawings which specify hole locations, outlines, conductor locations, thicknesses, materials, and surface finishes and inspection/quality requirements.

Chapter 5: Dielectric Materials

FPW dielectric systems are based on films and adhesives in which the films provide mechanical support and structure while the adhesives join conductor foil to film and coverlayer film to etched pattern. Commonly used films range from low-cost thermoplastics such as polyethylene terephthalate (PET) through epoxy-mat semiflexibles to adhesive-bonded and adhesiveless composites of metal and polyimide film. Several adhesive chemistries are used in systems which are thermoplastic, thermosetting, and hybrids of thermoplastic and thermoset.

The designer chooses the most cost-effective system by considering circuit size and complexity, environmental resistance required, termination technique—higher-temperature, higher-cost systems are needed for mass soldering—and expected circuit cost.

Adhesives have a profound effect on performance—more effect, in many cases, than the dielectric film—and should be used instead of the film type to categorize and characterize the systems. FPW adhesives are flexibilized and formulated to have low flow during lamination, a behavior which traces to bare-pad circuit designs and the necessity for preserving coverlayer openings.

Specifications and requirements for FPW materials—laminates, coverlayers, cast adhesive films—provide a universal means for defining and selecting performance against price, but do not control all important properties. Newer multilayer constructions, particularly rigid-flex circuitry, are better served by materials which are more like rigid PWB materials in thermal stress resistance. Most FPW is built in materials which were developed for single-layer, bare-pad designs but aren't suitable for these more demanding applications. A lower degree of flexibilization and higher cross-link density is needed in multilayer constructions.

Peel testing is the universal arbiter of laminate quality, but leads to wrong choices because it doesn't simulate a normal FPW failure mode and favors more plasticized, lower-temperature adhesives.

The coefficient of thermal expansion (CTE) is critically important in high-density multilayer constructions. Highly modified, flexibilized FPW adhesives show large CTEs, particularly at MIL-spec test temperatures; more cross-linked, rigid-like adhesives are more stable, with CTEs which are better matched to copper.

Potting and conformal coating are used to protect termination areas in FPW assemblies. Semiflexibilized systems minimize internal stresses in large cured sections and avoid stress concentrations where sharp bends occur near the FPW-potting compound interface.

Conformal coatings are thixotropic and hand-applied over solder joints or any sharp-edged, electrified hardware-FPW junctions.

Chapter 6: Conductive Materials

FPW conductor patterns are created by a wide variety of manufacturing processes ranging from subtractive print-and-etch through semiadditive techniques on thin foil seed layers to screen-printed polymer thick film (PTF) inking. The most common conductor material is rolled copper foil, but recent developments in electrodeposition technology have led to production of electrodeposited (ED) foils with exceptional flexural endurance which are ideal for fine-line, high-performance FPW use.

Other conductor foil materials include aluminum and low-conductivity alloys.

PTF conductors can't withstand soldering and therefore are assembled by zero insertion force (ZIF) connectors or conductive adhesives. Resistivity is high; current-carrying capacity is very low, but for nondemanding applications, PTF circuitry and associated specialized terminating techniques can save money.

Any circuit which isn't to be soldered can be built on lower-cost dielectrics such as PET, polyethylene napthanate (PEN), or vinyls, leading to important cost savings.

Shielding is a valuable feature in FPW; materials used include screening and etched copper foil, silver inks (PTF), and conductive tape wrappings. Efficiency is determined by conductivity and percentage of area covered; perforated shields increase flexibility but with reduced protection at higher frequencies.

Chapter 7: Adhesiveless Materials

Burdened by a functionally descriptive but not strictly accurate name, this new class of FPW laminates is expensive but boosts manufacturability and performance in high-layer-count circuitry more than enough to pay its way. Adhesiveless materials approach the stability and thermal resistance of fully cross-linked rigid dielectrics in a flexible laminate; this is achieved by eliminating conventional adhesives.

Three manufacturing methods are used:

- Coating a polymer precursor onto a foil
- Directly metallizing a polymer film
- Adhering foil and film together by means of a high-performance adhesive layer

Coating and metallizing methods produce composites in roll form; bonding method, because extreme temperatures are required, is only done in sheet form.

Major advantage of adhesiveless construction is improved dimensional and thermal stability. Because elevated temperatures aren't required, the deposited metallization process is most stable.

Deposited metallization produces any desired copper thickness and can provide PTH interconnection through vias at no added cost, making this production method ideal for MCM-L (multichip module interconnect based on laminates) and other high-density, fine-line circuit use.

Coating processes can produce very thin dielectric coatings on any foil (of handleable thickness), in any alloy.

The high-performance bond process can be applied to any polymer film and foil, but needs greater minimum thickness than the other methods.

The most significant product yield benefit of these materials appears in the MIL-P-50884 thermal stress test—exposure to 287.7°C for 10 seconds. Conventional adhesives, whose CTE is almost 20 times larger than that of copper, expand violently in the PTH-restricted coupon section, tearing the dielectric system apart. Adhesiveless materials, whose CTEs are less than 10 times copper, survive this test without failure.

Another important benefit is improved flexural endurance, which results from both reduced thickness (therefore force required to bend) as well as better foil support and firmer "centrality" of conductors within the cross section as it is flexed.

Secondary benefits are increased chemical resistance, better electrical performance, better dimensional stability, and reduced moisture absorption.

Peel strength of adhesiveless materials, because they do not contain extensible, flexibilized adhesives, is not as high as conventional materials. The mechanics of peel test discriminate against high-modulus, stiff bonds leading to erroneous conclusions about product quality.

Chapter 8: Manufacturing Processes

This chapter details the many process steps, the equipment, and engineering considerations which are involved in volume production of FPW.

Overall methodology

Roll-to-roll processing is used for high-volume production of single-sided circuitry; panel processing is the choice for special or multiple-process, smaller-lot, quick-turn production. Roll-to-roll processing generates circuitry at the lowest cost, with tight process controls, but can't be used economically for short runs, and can't economically be tooled to provide all the myriad variations which FPW requires.

Materials

The most important FPW manufacturing issue is dimensional stability, which determines maximum panel size, thus labor per circuit. Smaller panels provide tighter registration of tool to part for higher precision; more stable materials allow larger panels, thereby lowering material cost (because of more efficient circuit nesting) and increasing yield.

Coverlayering is a distinct FPW characteristic; this is application of another dielectric layer onto the etched pattern to provide full insulation and protection of the conductor pattern from handling damage or accidental short-circuiting. Coverlayering requires preparation of the layer by perforating the desired access openings followed by lamination, which brings in the stability/alignment issue.

If the coverlayer is applied as a full sheet without apertures, alignment isn't required, but baring pads after lamination is a difficult task. A few openings can be cost-effectively machined or etched if this avoids the labor of preparing a coverlayer. However, since the cost of machining openings after lamination depends on the number of openings, it's not cost-effective if many are required; preperforation then is the cheapest technique in spite of alignment issues and steps which are required to minimize adhesive flow onto the pads.

A surface finish on exposed pads or PTH terminations is normally required. A common finish is fused tin-lead plating, which aids solder assembly. Other finishes include precious metals and organic protective (antioxidant) coatings.

Lamination

Production of base laminates and application of coverlayers is performed in laminating presses or autoclaves. The press process is lower in cost and faster and provides high pressures and temperatures, but is influenced by variations in material thickness. Press pads are used to accommodate or neutralize variations in product thickness. Autoclave equipment is more costly and the process is slower; it cannot reach the higher pressures which are needed for bonding high-viscosity systems but provides more uniform pressure provided that alignment fixtures are well-maintained. Both press and autoclave lamination use the vacuum bag process to improve extraction of volatiles and to minimize voids in a cured structure.

The best lamination equipment is the vacuum press, a hydraulic press enclosed in a vacuum chamber, which eliminates the labor and material cost of vacuum bagging. Vacuum presses also provide better extraction of volatiles from the press load, since the edges of the load are fully exposed to the chamber.

Imaging

Silk-screen resist imaging is useful in coarse circuit designs and PTF process, but photoimaging is the most prevalent technique. PTF requires a screening process because conductivity is a direct function of image thickness, and silk-screening provides maximum thickness. Photoprocessing utilizes a resist coating which is applied as a laminated "dry" film, or by deposition from a liquid bath.

The photoresist process can be either negative or positive, depending on whether the exposed and developed coating is a reversed or direct copy of the artwork. Likewise, resist images can either protect the desired areas—etching images—or expose them—plating images—depending on intended conductor-generating process.

Resist thickness directly controls resolution; liquid-applied coatings, which measure 0.0005 in or less, can resolve lines and spaces on the order of 0.001 in in size.

The dry film process is popular because it is forgiving and rapid and has adequate performance for most current designs. Dry film, because the resist image is thicker, is used for semiadditive production in which conductor patterns are electrodeposited in the image: resist sidewalls (0.001 in or more) allow production of equally thick conductors with photo precision.

Tenting—the ability to protect PTH—is an advantage of dry film resists. Electrophoretically deposited liquid resists can also coat and protect the walls of these holes.

Photoresist development takes place in conveyerized equipment that applies mild caustic chemicals by spraying. Similar equipment, but with slightly more concentrated chemistry, is used to remove, or strip, the resist from etched panels.

The metal resist process is common in PTH production. This is a continuation of the process of imaging for pattern plating (conductor areas are exposed by the image and built up to thickness by electrodeposition); the plated conductor pattern is overplated with a layer of tin-lead before the resist image is removed. Ammoniated cupric chloride does not attack tin-lead very rapidly, therefore the overplate, which is applied everywhere on the conductor pattern including through the PTH, protects these areas as the background or seed foil cladding is removed. At the end of the process, the tin-lead coating is fused to convert it into a solder coating for improved solder assembly and long-term solderability, or stripped off in cases which require base copper circuitry.

Specialized imaging methods range from die stamping for very high-volume, coarse circuitry to electroforming.

A specialized technique for generating conductor patterns, electroforming isn't an imaging process in the standard sense. This process consists of plating up the conductor pattern on a permanently patterned mandrel, then transfer-laminating it onto a dielectric system to form the finished circuit. The mandrel is indefinitely reusable. This method allows production of precious-metal-coated and flush-laminated circuitry because it proceeds in the reverse direction, compared to ordinary production methods.

Etching

Etching—the process of dissolving unwanted foil areas—involves several different chemistries, of which the two most popular are ammoniacal cupric chloride and acidic cupric chloride.

These and other etching baths are monitored and maintained at constant activity levels by additions or by feed-and-bleed systems which introduce fresh etchant and pump spent bath into holding tanks for removal.

Etchers are conveyerized equipment similar to those used for resist development and strip. Etching is accomplished by spray nozzles which uniformly and powerfully apply the etchant to both surfaces of the circuit panels as they are conveyed through the machine.

Perpendicularity of the etched sidewalls is an important process consideration. Better chemistries and equipment can provide performance which duplicates artwork dimensions within about 0.001 in per 0.0014 in of copper thickness. This so-called etch factor is used to correct the artwork so that the finished conductors are the desired size.

Controlled variables in etching consist of conveyer speed, spray pressure, bath temperature, and composition.

A specialized form of reversed electroplating called *electroetching* can be used in some situations. The process is similar to electroforming, but done backward.

Stripping

Resist images are removed by conveyerized equipment similar to etchers or developers; the process chemistry, a slightly stronger and warmer form of developer, is sprayed onto the panels from both sides.

Outlining

Cutting the outline contours of FPW is usually performed with low-cost knife-edged dies called *steel rule dies* (SRDs). These are produced semiautomatically from computer-aided design (CAD) data generated at the design stage. SRDs are aligned to the FPW by means of tooling pins which fit into holes punched or drilled in the FPW panel; the die is forced through the circuit by a hydraulic press.

Alternative techniques include matched-die cutting (used for precision or long life), routing, and use of scissors and hand knives.

Tooling

FPW is a custom product and most production tooling is also custom-designed and custom-produced for each circuit pattern. A common concern with all FPW tools is alignment to the circuit pattern, which is degraded by the dimensional instability and inherent weakness of FPW laminates.

The best tool alignment means is optical, direct to the circuit pattern by robotic self-aligners. An example is the postetch-punch (PEP) process for which a series of fiduciary marks are included in the artwork. Automated equipment, if brought to rough alignment with these marks, grips and moves the panel into precise registration, then punches an accurately centered tooling hole in the fiduciary mark. Although this method cannot totally neutralize material shrinkage, it does reduce its effect since multiple PEP-generated holes are used to align each tool, therefore averaging of hole centers improves precision.

Process flows

A wide range of process flows are used to produce even the simplest single-layer FPW. These include upside-down, reverse bare, PTF, and semiadditive methods. Each process has advantages and disadvantages.

Best concentricity of throughhole to pad to coverlayer occurs with the upside-down process, because all three features are applied to the unetched copper foil sheet. However, all three features then suffer positional shift as the etched panel shrinks.

The choice of process flow is based on which feature should be best centered, specifics of design such as how many apertures are on either side of the panel, production volume, and shop custom.

The PTH process is a powerful technique for increasing connectivity by using the third dimension to route conductors past obstructions—connector patterns are a good example—or to allow conductor position interchange to satisfy pin-to-pin interconnection patterns.

PTH requires careful alignment and registration, because multiple conductor layers are involved, all of which must be aligned to each other, to the hole pattern, and to pad-layer artwork. Additional special processing is needed to clean dielectric residues out of the drilled holes and for electroless deposition processes which make the hole walls electrically conductive with a strongly adherent deposit.

The PTH process can be practiced in panel, pattern, selective, or semiadditive variations depending on conductor density and manufacturing preference.

Chapter 9: Rigid-Flex Manufacture

Rigid-flex circuitry is a complicated combination of rigid and flexible dielectrics in a multilayer, PTH-interconnected circuit. Outer layers in the rigid areas are PWB laminate materials which protect the inner FPW layers from assembly stress and handling damage. FPW layers extend beyond and interconnect between rigid areas.

Design and manufacture of RF circuitry is not much different from that of multilayer FPW circuitry. Maintenance of layer alignment and dielectric integrity with increasing layer count are the primary production concerns.

If MIL-P-50884 acceptance testing is required, yield can be seriously affected by excessive thermal expansion of FPW materials. These materials, as used in conventional FPW constructions, have enormous CTEs at military-acceptance temperatures.

Process steps

Procedures for preparing coverlayered FPW layers and producing multilayered PWB cap boards which protect the rigid areas are straightforward. Added process steps which are required for RF production are directed to the generation of the circuit outline and to venting and sealing the laminated panel for wet processing.

Outlining

Because there are mixed dielectrics in varying thicknesses, the outline is cut in several separate process steps:

1. Edges of FPW layers extending between rigid areas are partially outlined or slit in layer form.
2. Rigid areas are also partly outlined or slotted, to predefine the rigid edge which overlies the exit point of the FPW layers.
3. The final outline is cut by routing from slot edge to slot edge; this cut meets the precut edges of the FPW layers at the rigid-area outline, thus fully defining the circuit.

Venting and sealing

RF circuitry panels must be completely sealed against intrusion of wet process chemicals which are used in the PTH process. This requires a complicated assembly of materials, called a *bag,* which temporarily seals the slot areas.

Because some of the interior of a laminated RF panel is unbonded, venting is necessary to relieve unbalanced pressures when the panel is exposed to vacuum during the plasma process. Vent holes are left open during plasma processing, then sealed before wet processing.

Bookbinder design

Because RF circuits are primarily used in dense packages, bookbinder design, or progressed FPW layer lengths, is sometimes employed to reduce installed stress. This design forces additional attention and tooling to accommodate increased panel height during lamination,

drill and rout, resist cladding, and assembly operat
protruding FPW layers for wet processing is another

Materials

Simple RF circuits can be built in conventional PWL ....
rials. Above eight layers, where pattern density is high, or when MIL-P-50884 qualification is required, production yield drops, and use of premium-cost adhesiveless laminates becomes cost-effective. These developing materials have lower CTEs, better dimensional stability, improved chemical resistance, and reduced moisture absorption.

Conventional FPW adhesives can be used to bond together FPW layers in RF construction, but their reduced flow properties and poor thermal stability result in high adhesive content and failure at acceptance test. Rigid prepregs are superior for bonding RF constructions and are used extensively for this purpose.

Flow control

Increased flow during lamination is desirable in producing solidly bonded RF panels with minimum adhesive thickness, but flow into slot areas or between FPW layers is undesirable. Careful attention to material properties and lamination protocols plus use of correct filler materials minimize flow beyond rigid area boundaries.

Chapter 10: Standards and Specifications

FPW is a product of military need, and heavy inspection and quality control activity is the norm. Complete sets of standards for materials, process, quality system, and product are available from the military services and the Institute for Packaging and Interconnecting Electronic Circuits (IPC), a major industry association.

Visual inspection is commonly used at all stages of FPW production, including after acceptance and at customer inventories. Transparent dielectrics and thin sections make FPW very accessible to visual study.

Dimensional evaluation is difficult in FPW because the product is not naturally flat; it must be restrained for measurement. Cluster dimensioning with tight tolerances only within hole patterns that are generated by hard tooling and fractional or minimum dimensions between is appropriate.

Electrical testing is preferred because it yields numeric results and represents the function of the product.

Careful review of FPW data packages is needed to assure that double dimensioning between mechanical requirements and supplied artrk data doesn't occur. Other likely areas of conflict: specified conductor resistances, capacities, or impedance.

Chapter 11: Assembly

Termination

FPW is joined to other components primarily by soldering. Either hand or mass termination techniques can be used. Mass termination includes wave and reflow soldering; hot-air ovens are widely used because they provide accurate temperature control, shockless heat-up and gentle cool-down.

Mass technique with good masking allows solder attachment to FPW which has low-temperature (therefore low-cost) insulation.

Pressure termination is used. Techniques include crimping, zero or low insertion force (ZIF or LIF) connectors, direct pressure connectors, and screw terminations.

Conductive adhesives are heavily used with PTF circuitry and provide low-cost mass-termination ability, but with limited conductivity and current-carrying capacity.

Surface-mount assembly

Because component size is small and growing smaller, direct attachment to FPW lands is practical and is frequently used to produce extremely dense, lightweight, compact packages. Automated pick-and-place equipment with hot-air reflow provides high throughput with small investment in capital and space.

Forming

FPW can be shaped by use of jigs and fixtures. Polyimide dielectrics are difficult to form and may delaminate or wrinkle under sustained stress and elevated temperatures.

Inclusion of extra conductor foil at the bend line aids in accurate forming.

Potting, conformal coating

Application of postassembly insulation coatings to restore circuit insulation and protect other parts of the FPW from scratching by sharp edges is commonly practiced. Total encapsulation or potting requires molds and careful technique; overcoating with thixotropic materials—conformal coating—is easier and almost as beneficial.

Nomenclature

When space for built-in legends, part numbers, terminal identification, and date codes isn't available, this information is stamped or printed on the exterior surface (or onto a label which is applied to the surface). So-called nomenclature is important for assuring correct assembly, as an aid to field service and repair, and for quality traceability.

Mechanical

Stiffeners, connector bodies, and attachment hardware are fastened to FPW by a variety of techniques which include adhesives, rivets and eyelets, bolts, and screws.

Chapter 12: Examples of High-Volume and Unusual Flexible Printed Wiring

This chapter describes:

Military reel cable. 50 ft of polyester-polyethylene FPW, manufactured by the fusion process with silk-screen printed image.

High-volume epoxy mat FPW. A telecommunications circuit manufactured in high volumes by the panel process, using punched PTH holes to interconnect the double-sided design. The dielectric is epoxy-mat with tin-lead metal-etch resist and permanent surface finish.

High-volume vinyl FPW. A telecommunications circuit manufactured by a semi–roll-to-roll process on vinyl-copper laminate. Coverlayered with vinyl film by the fusion process. The surface finish is electroless tin.

Stripline. A 10-ft-long three-layer, eight-conductor 93-Ω transmission cable for computer use. Built by a combination of roll lamination and press lamination using 0.008-$\times$0.025-in copper conductors and a novel quilting process to improve flexibility in a foam dielectric system.

High-voltage ignitor cable. A two-layer FPW with welded and soldered termination hardware. Made by the fusion lamination process with Kel-F dielectric to produce high-voltage (5000-V dc test) ignitor circuits.

Vacuumtight stripline. Solid dielectric construction with high-pressure lamination produced a 50-Ω multiconductor FPW cable suitable for use as a feedthrough in vacuum instrumentation.

Outsized backplane. Eighteen-layer, 4.4-m (14-ft, 8-in) backplane for scientific instrument use. Contains 336 signal conductors at 35-Ω impedance in eight shielded conductor layers. Interleaved shields are interconnected by edge painting with silver conductive ink.

High-volume, miniature multilayer backplanes. A compact, solder-interconnected design containing up to 12 signal layers sandwiched between $\frac{1}{32}$-in epoxy-glass boards. Layers are not bonded and are solder-attached to interconnecting copper eyelets which clamp the top and bottom boards together in the finished circuit.

Appendix A

Foreign Material

This topic is discussed briefly in Chap. 3, in sections of Chap. 8, and other sections of this book.

FPW is uniquely vulnerable to rejection for foreign material (FM) inclusions—unidentifiable particles permanently bonded between layers of dielectric or adhesive—because it's fabricated from multiple layers of transparent dielectric films (and prepregs, in the case of RF), each surface of which can collect contamination, and because it's exposed to elevated process temperatures which blacken and expand submicroscopic particles up to a size which is detectable. FM is composed of small particles from a wide range of sources, always present in enormous quantities in a typical FPW factory, which are attracted and bonded to the dielectric films by static charges and stubbornly resist all but the most sophisticated removal methods. Most FM is introduced during FPW lamination processes, but it should be realized that purchased dielectric films and adhesive coatings can include built-in FM as well. The FPW manufacturer is advised to inspect incoming material at least as closely as his final product. Most quality standards reject FPW with FM between conductor runs and make no provision for rework or repair.

FM, defined as anything other than dielectric or conductor metal, is typically organic in nature—hairs, clothing fibers, food residues—and, because embedded inside the FPW, unidentifiable and irremovable. Since it's impossible to probe and verify long-term inertness of these particles, prudence forces rejection as the safe action.

Handling, machining and conventional cleaning of coverlayers, bondplies and etched inner layers of multilayer constructions generate FM particles and increase static charges; all should be minimized. Wiping or brushing techniques used in preparation are particularly counter-productive in removing FM because rubbing or stroking or separating one sheet from another generates greater static charges which increase the unwanted attractive forces. The best strategy is to prepare etched layers and dielectric sheets for lamination in the most economically practical clean environment: i.e., laminar-flow cabinets and cleanroom conditions (smocks, hairnets, etc; see Fig. 8-5) and to neutralize static charges by contacting the surfaces with a conductive medium, then to displace and capture the FM. These methods include:

- wriststraps and conductive mats
- ionized air steams
- conductive bristle brushes
- plasma or corona treatment
- wiping with conductive-liquid moistened pads to dissipate static electricity
- tacky rollers
- vacuuming to capture and remove the particles

Appendix

B

Adhesiveless Materials in Dewar Circuitry

When FPW is employed as the interconnecting technique for IR sensors which are cooled to cryogenic temperatures to enhance their sensitivity, a significant new design requirement—reduced thermal conduction—comes into play. In conventional applications, FPW is designed to meet low resistance requirements, in Dewar use high resistance to thermal conduction is desired to reduce the thermal load on the cryogenic system which house the sensors.

Conductivity, electrical, or thermal, is a function of the crossectional area and physical properties of the conducting metal. In traditional laminated materials which include 1oz copper foils, the minimum producible conductor width—003"—is far too wide to yield the desired low thermal conduction because copper is an excellent conductor of electricity and thermal energy.

Minimum width of an etched run is primarily determined by foil thickness, because of undercut and etch factor effects and breakdown of the resist image with prolonged etch cycles. For this reason, the technique for building cryogenic FPW has been to use thin layers (.0005") of very high resistance alloys—Constant and Balco, for example, 30 to 40 times higher in resistivity than copper—which are overplated with 100 to 200 microinches of copper to create the desired conductivity. This is a complicated and expensive manufacturing technology which forces the FPW maker to procure and laminate expensive, delicate alloys onto an FPW dielectric systems, then to plate it with a well-adhered but thin and uniform layer of copper, and finally to etch this duplex metal into a fine-line, tight-toleranced FPW.

Deposited metalization adhesiveless materials are commercially available with 100 to 200 microinches of copper, thus eliminating all exotic materials and special processing. These thin-clad adhesiveless materials allow relatively easy production of conductors as narrow as the photoresist can define: with liquid systems, that's .001" or less. Thus adhesiveless materials provide a single-metal, simplified process for production of Fpw which has suitable levels of thermal and electrical conductivity for Dewar applications.

Glossary

A zone In rigid-flex circuitry, the cross-sectional area contiguous with a PTH.

Activation The process of chemically treating a nonconductive surface to make it receptive to metallic deposition.

Additive The process of building up a conductor pattern by selective plating or other addition technique on an initially nonconductive substrate. Contrast with *semiadditive processing* and *subtractive etching*; see *seed layer*.

Adhesive A composition, usually polymeric in nature, which is used to permanently join or bond materials together.

Adhesiveless Applied to a class of flexible composites consisting of dielectric and foil without discernible adhesive.

Aligning; alignment Process of locating two or more features or objects in the correct relative position.

Annular ring The exposed conductive area surrounding a central hole usually in a terminal pad.

Aperture An opening or window in a dielectric layer which provides access or contact to an underlying conductor or pad.

Artwork A dimensionally precise photographic template of a circuit pattern. Used with photomechanical processes.

B zone In rigid-flex circuitry, the cross-sectional region between PTH.

Bag layers In rigid-flex circuitry, added layers whose purpose is to seal off slots and other premachined cap board apertures.

Balanced construction Applied to FPW in which the base and coverlayer dielectrics are identical in composition and thickness and the foil layer is centered in the cross section. Preferred for high flexural endurance.

Baring; reverse baring; full baring The process of removing a dielectric coating (see *coverlayer*) to expose a conductor or pad. Reverse baring is the process of removing the base dielectric. Full baring is the process of removing both coverlayer and base dielectrics.

Base layer; base dielectric   The layer of dielectric which supports a conductor pattern. May be the dielectric portion of the flexible laminate or the dielectric layer which receives conductive ink patterns. Combined with a very thin conductive deposition (see *seed layer*) forms the starting material for semiadditive process.

Beading   A smoothly radiused fillet of resilient material which is applied to the junction between rigid and flexible areas in rigid-flex to distribute bending stresses and protect against chafing.

Bondply   A film dielectric layer with adhesive on both surfaces; used to join together or bond layers of uncoverlayered FPW.

Bookbinder   An RF design which utilizes progression in FPW.

Busbar   A heavy interconnecting circuit which joins several other circuits together; usually for plating or power distribution.

CAD/CAM   Computer-aided design and computer-aided manufacture; a system of equipment, software, and management philosophy for maximizing automation in the preparation of artwork and tooling, including NC data.

Cap boards   Outermost, rigid-laminate layers in rigid-flex.

Cap lamination   Use of single-sided PWB laminate as cap boards in a multilayer structure.

Cast adhesive   A layer of unsupported adhesive, normally on a temporary carrier, used to bond together coverlayered FPW.

Circuit pattern   The set of conductors, nomenclature, and pads which forms an interconnect image. See *artwork*.

Clad foil   A foil layer which is bonded to a base layer of dielectric to form a base laminate. To be distinguished from plated layers which are applied over the clad foil in PTH process.

Cluster registration   A technique for nullifying overall dimensional change by use of local points of origin for measurements and tool alignment.

Composite   A multiple, efficiently compacted array of circuit patterns organized for best production yield from a panel.

Concentric   Having a common center, as when the central hole, pad outline, and covercoat opening are concentric.

Conductive ink   A composition of binder and conductive particles which is printed on a base dielectric layer to form a conductive pattern.

Conductor layer   The layer within a flexible circuit which contains the conductor pattern; typically centered between base and coverlayers of dielectric.

Controlled impedance   An electronic concept relating to high-frequency circuitry in which the distributed inductance and capacitance of a conductor are designed for efficient transmission of signal pulses.

Coupon   A small test circuit included in a panel design.

Covercoat   A layer of dielectric with openings or apertures which insulates the conductor pattern while providing access to selected locations. Applied as a liquid or film over an imaged conductor pattern, but in the case of upside-down fabrication may be the starting layer of dielectric. See *coverlayer*.

Covercoat opening   An opening which provides access to the conductor pattern or pad (see *aperture, baring*).

Coverlayer   A film of dielectric material with adhesive, usually identical with the base layer, which is bonded over the etched conductor runs to insulate them.

CTE   Coefficient of thermal expansion.

Date code   A mark which defines the manufacturing date of a product.

Detail   A layer or subassembly in a multilayer product; the details are laminated together to form a multilayer circuit.

Dielectric   A nonconductor of current.

Differential etch   The process of removing the seed layer from a semiadditive circuit by exposure of the unresisted circuit to a brief etching cycle. Both conductor pattern and seed layer are attacked; the conductor pattern remains after the process because of its greater thickness.

Dimensional stability   Ability of a laminate to maintain desired size through exposure to process stress (temperature, pressure, humidity, etc.). Commonly specified in two orthogonal axes, i.e., MD (machine direction, parallel to the roll length) and TD (transverse direction, across the roll width).

Distortion   In etched patterns, irregular or inconsistent deviation from expected size; nonuniform shrinkage or expansion.

Double-sided bare FPW   FPW with apertures through base and coverlayers.

Double-sided FPW   FPW having two conductor layers.

Double-treat   A copper foil with an adhesion-promoting and passivating treatment on both surfaces.

Etch; etching   Process of removing material to form a desired condition or pattern. Usually accomplished by chemical solution.

Etch factor   Ratio of foil thickness to per-side reduction in feature width.

Etched foil conductors   Conductors formed from a sheet of foil by an etchant resist and etching process.

Eyeball alignment or **registration**   Use of operator vision to align materials or tools. Includes the concept of optimization or best fit between multiple fiduciaries.

Factoring   The process of adjusting the scale or proportionality between artwork and desired circuit size. The adjustment is usually stated in thousandths of an inch per inch of expansion, e.g., 0.9, 1.3.

Fiduciary   A nonelectrical reference artwork mark or feature which locates the center of a hole or measurement grid.

Fillers   Sections of release material inlaid into windows of multilayer details to restore thickness and inhibit distortion.

Fine pitch   Applied to a pattern of pads or contacts which are located center to center at a distance, or pitch, which is small enough to affect the process.

First article   The proof-of-process circuit, first manufactured after process or tooling change; used to verify process and tooling.

Foil   Metal in sheet form but less than 0.007 in thick.

Foil lamination   Use of foil with prepreg to create a cap board at lamination in a multilayer assembly.

FPC   Flexible printed circuit. A reproducible pattern of conductors with components supported by or encapsulated in a flexible dielectric system.

FPW   Flexible printed wiring. A reproducible pattern of conductors supported by or encapsulated in a flexible dielectric system; the interconnect part of an FPC.

Fully bared   See *baring*.

Ground plane   A large area of conductive material which serves as a reference point for voltage distribution or measurement and/or as a shield.

Hash marks   Features designed into artwork to define the location of folds, stiffeners, and the like; alignment guides.

HASL   Hot-air solder leveling, a process for applying a thin coating of solder to FPW and PWB.

IPC   The Institute for Interconnecting and Packaging Electronic Circuits, a leading trade association of FPW manufacturers.

Keying   Process of fitting together in a unique, desired relationship.

Laminate   (Noun) A composite consisting of dielectric and conductor layers; (Verb) The process of joining layers of materials by application of heat and pressure.

Laminate voids   Rejectable voids, bubbles, or holes in PTH cross-sectional coupons.

Land   (Usually) An enlarged section of a conductor intended for contact or termination.

LIF   Low insertion force; applied to a connector which has mild contact force at insertion and a mechanism to increase contact normal force after insertion.

Marking   Application of nomenclature.

Microstrip   A controlled-impedance design consisting of a signal conductor spaced apart from a reference plane by a dielectric. See *stripline*.

Multilayer   A PWB or FPW with more than two conductor layers.

NC   Numerical control.

Nesting   Process of arranging circuit repeats for best use of area in a composite; packing them together closely.

Nomenclature Nonelectrical parts of a conductor pattern. Usually includes part numbers, identification of contact points, etc. The process of applying nomenclature is called *marking*.

OPC Organic protective coating, a coating which is applied to pads to preserve their solderability in storage.

Outline The desired edge contours or borders of FPW or PWB; to cut a circuit from its panel along desired border contours.

Pad See *land*.

Palletized A manufacturing technique in which completed, 99% outlined circuits remain in panel form for assembly convenience.

Panel A sheet of laminate of standard size containing a group of circuit patterns. Size is determined by best manufacturing efficiency (distortion and shrinkage are major factors) as well as circuit size and material utilization. Common FPW sizes are 12×16 in, 12×18 in, and 18×24 in.

Panelized See *palletized*.

Peel strength The strength of the bond between two sheet materials, expressed as the force required to peel them apart.

Postetch punch (PEP) The technique of generating tooling holes by reference to etched fiduciaries.

Pouch An intentionally unbonded area within a laminated panel of rigid-flex circuits; usually the FPW area.

Prepreg Glass cloth or other saturating mat coated with B stage resin; used as a bondply or layer in PWB and rigid-flex constructions.

Press pad An assembly of layers of material used during a lamination process to distribute pressure as desired. Includes a release, or nonadherable, surface with crushable or hydraulic conforming layers.

PTF Polymer thick film.

PTH Plated throughhole; an interconnecting hole which passes through internal and external pads and layers to be interconnected, whose walls are made conductive by deposition of metallic plating.

PWB Printed wiring board.

Registration Alignment between two or more features; the process of aligning features.

Release film or agent An unbondable material.

Repeats The number of circuit patterns in a composite. See *up*.

Resist A material which protects an underlying layer from a process; in photosensitive resists there are two forms: *negative-acting* (the area exposed by the artwork remains after development) and *positive-acting* (area exposed by the artwork is removed by development).

Resolution Resolving power; ability of a material or process to clearly define closely spaced features.

Return to panel A palletizing process in which a circuit is fully outlined, then pressed back into the panel for automated component assembly.

Reverse baring See baring.

RF Rigid-flex.

Roll to roll A method of manufacture utilizing long rolls of material which pass continuously through production equipment. Requires large capital investment and careful engineering but gives best process control, least handling damage, and lowest cost (for simple circuitry). Contrast with *panel.*

Run list Electrical interconnect pattern in which all runs to a given termination are grouped together; a from-to list.

Screen printing A printing process in which the ink is forced through a stencil, or screen, onto a surface by a squeegee.

Seed layer A very thin but conductive layer used in semiadditive process. See *additive, subtractive etching, semiadditive processing.*

Selective plating A process for manufacturing FPW with PTH terminations; PTH plating is confined to the termination pads and holes.

Semiadditive processing Creating a conductor pattern by electroplating onto a seed layer through a resist image.

Sequential lamination A process in RF manufacture for providing PTH termination in rigid areas of differing thickness.

Single-sided FPW FPW with a single conductor layer; contrast with *double-sided* and *multilayer.*

Slot rout Process of machining the edge of a cap board, prior to lamination into a rigid-flex panel, to define the FPW edge.

Soda straws Extended voids along the root of etched conductors caused by inadequate coverlayer adhesive flow.

Steel rule die A low-cost tool used for outlining or punching; consists of a pattern of sharp-edged blades, similar to steel rulers, embedded edgewise in a die block.

Step and repeat A technique for generating composites by uniformly repeating the circuit pattern at predetermined intervals in both directions. Contrast with *composite,* an arrangement of patterns which are nested together for greatest yield but not necessarily at uniform spacing.

Stripline A controlled-impedance design consisting of a central signal conductor between overlying and underlying shield planes.

Substrate The support layer, usually a dielectric material, on which a process step is performed.

Subtractive etching The process of creating conductor patterns by removing unwanted areas of a foil layer.

Surface conductors Conductors on the outer surface of cap boards.

Surface finish The final treatment applied to pads and land areas to preserve their suitability for subsequent use.

Tack To temporarily secure in position, usually by local heat and pressure or solvent action on an adhesive coating.

Tent To form a film of etchant resist which protects holes from etchant. A resist is said to *tent* when it bridges across throughholes.

Terminal See *land*.

Thermal shock A test procedure consisting of rapid cycling from low to high temperatures.

Thermal stress A test procedure consisting of conditioning, then floating a coupon on solder, followed by examination including cross-sectioning.

Throughhole An opening or aperture which penetrates entirely through an FPW, PWB, and RF to provide access from top to bottom surface, and by PTH process (if used) to any intersecting internal lands.

Tinning Process of coating with hot solder.

Tool A device which reduces labor or improves precision; to engineer a process or equipment for efficient production.

Tooling hole A nonelectrical hole which serves to align or register a tool or process to a circuit or panel (see *fiduciary*).

Trim lines Artwork or drawing lines which define the location of cuts.

Tugs Small sections of the outline of a circuit which are intentionally uncut at the outline stage to keep the circuit in its panel until later in the process.

Undercutting Penetration of etchant under a resist image resulting in narrower features.

Unsupported A land or pad which has no dielectric support on either surface.

Up The number of circuit repeats in a composite, as in *three-up*.

Vacuum lamination Process of lamination in a vacuum to reduce trapped volatiles. The vacuum environment is created either by enclosing the press load in a turkey bag which is then evacuated or by using a press whose platens are enclosed in a chamber (eliminating the labor and cost of bagging and debagging).

Vent holes Small holes drilled through a cap board and into a pouch area to relieve pressure buildup.

ZIF Zero insertion force; applied to a connector which includes a camming device to apply all normal force after the mating part has been inserted.

Index

INDEX NOTE: The *f.* after a page number refers to a figure.

Additive process, 143, 149
Adhesive layer, 85
Adhesiveless (materials), 4, 6, 19, 33, 90, 128
  benefits, 117
  constructions, 114–117
  definition, 113–114
  in RF, 205–212
  tables, 119
    CTE effects, 208–212
    thickness, 207
  in forming, 164
  in solder wicking, 135
Adhesives, 4, 14, 58, 86, 133–134
  flow, 90, 100, 129, 131–134, 212
  in RF, 207, 212–214
  types, 92
Alignment:
  eyeball, 7
  tooling holes, 7, 46, 49, 50, 65, 66, 129, 164
Amide-imide adhesiveless, 115
Annular lands, 41, 129
  effect on cost, 52
  in PTH, 44
Aperature; in coverlayer, 3, 52, 128
  post-lamination, 135
Aramid paper (Nomex), 95
Artwork, 5, 45, 46, 162, 220
  factoring, 22, 69, 80
  stability, 81
Aspect ratio, 50
Assembly methods, 86
  cost, 12, 51
Automation; automation devices, 126, 147

Back-to-back process, 26

Bag process (RF), 196
Bagging (for lamination) ("turkey bags"), 141–142
Banking agents (in etch chemistry), 158
Bond strength, 86
Bondply, 91
Bookbinder, 35, 189, 195, 198
  (*See also* Progression)
Borders, 50, 68, 193
  effect on stability, 68–169
Butyral phenolic adhesives, 2, 92

Cap boards:
  in RF, 205
Cast adhesives, 85, 91
Chemical resistance, 86
Cluster alignment:
  multipe datums, 7, 22, 165
Coefficient of thermal expansion (CTE), 22, 90, 93, 114, 124, 208
Composites, 7, 49, 66, 68
Computer aided design (CAD), 162, 220
Conductive adhesive, 6, 38, 151, 236
Conformal coat, 53, 85, 102, 240
Connectors, 40, 52, 212–216, 239
  pressure, 232–235
Controlled impedance, 73–80, 247, 249
  design, 75
Cost, 10–11, 21, 51, 66
Coupons, 64, 68, 167*f.*, 188, 221
Covercoat:
  covercoating, 2, 52–53, 128, 136
  photoimageable, 121
Coverlayer:
  coverlayering, 2, 28, 33, 38, 46–47, 52, 68, 85
  lamination, 128–130

287

Cracking (in foils), 26
Current capacity, 13
  chart, 32–33

Dahlgren, Victor, 2
Data requirements in design, 30
Definitions of FPW components, 5
Deposited metallization in adhesiveless material, 115
Desiccation, 140, 213
Design rules, 32
Development (of images), 147
Die cut tolerances, 160
Die stamp (for imaging), 151, 173
Dielectric breakdown in design, 34
Dielectrics, 85
  strength, 15
  effect on cost, 38, 51
Dimensional stability:
  effect on yield, 3, 20–22, 65, 68, 95, 113, 136, 172
Documentation, 80
Double-dimensioning, 23, 82, 136, 219
Drilling, 129–131
  in PTH, 176
Dynamics, 13, 53–54, 55, 57

Electrodeposition (ED), 27, 104
Electroforming, 150
Electroless deposition:
  for PTH, 178
  surface finish, 137
Environment, 14, 15
Etch, 152
  chemistry, 155
  electric, 155
  equipment, 152–155
  immersion, 154
  planetary, 157
Etch chemistry, 155–156
Etch factor, 33
Exposure, 146–147

Factor (artwork), 69
Fiduciaries, 7, 126, 147
Films, 86, 93
  chart, 97
  descriptions, 000
Fixtures (listing of), 47

Foils, 103
  chart, 106, 110
  copper (ED), (RA), 103, 106
  LTA, 105–106
  Special alloys, 109
Flammability, 86, 92
  in adhesives, 14, 86, 96, 99
Flash plate, 181
Flexible printed wiring, 4–5
  benefits, 17
  definition of, 9
  elements, 85
Flexural endurance, 33
  chart, 107
  foil effects, 105
  materials effects, 16, 52–57
Foreign material, 91
Forming, 14, 15, 57, 163, 212, 229, 236
Freedom of pin address:
  benefits in design, 36
Fusion bond, 3, 4, 24, 87, 95, 245, 248–249

Handling, 13, 24
  back-to-back process, 26
  foil effect on, 26
  in lamination, 140
Hardware, 101, 225
History, 2
Hitch-feed:
  in lamination, 87
Hot air solder level (HASL), 8, 138

Imaging:
  alternate methods, 172
  positive negative, subtractive-additive, 143–144
Inspection, 12, 50
  transparency:
    effect of, 20

Kel-F, 2, 87, 249
Keying, 10

LTA (low temperature anneal), foil, 106
Laminates:
  adhesive, 4, 127–128

Lamination, 131–142
    autoclave, 142
    panel process, 139
    roll-to-roll, 138
    vac assisted, 141
Leaders, 24–125, 154
Low flow:
    in adhesives, 90

MCM-L, 9, 117
Machining in RF:
    rout; window; slot, 199
Materials, 2, 204
    effect on cost, 51
    effect on design, 52
    QC for, 127

NASA, 225
Nomenclature, 8, 63, 68, 201, 241

Organic protective coat, 8, 138, 222
Ounce (obsolete foil thickness metric), 107
Outline, 7, 46, 50, 68, 82, 160

PTF (polymer thick film), 6, 72, 86, 110, 151, 173
PTH, 7, 71–72, 90, 136, 140, 172, 247
    design for, 35, 41, 44, 145
Packing; in design (nesting), 31, 35, 49, 66
Panelization; palletization, 60, 201
Panel process, 22, 125, 126
    cost effects, 66
Paper dolls (design aid), 34, 66
Patents; of Sanders Associates, 3
Peel strength, 90, 208
    compared to tensile, 123, 143
Photoresist, 143–148
    tenting, 145–149
    warnings, 147–149
Pinning, 164
Plasma process, 177, 203
Plasticization, 6, 122
    effect on peel strength, 122
    effect on thermal props, 208
Polyester, 6, 15, 51, 86, 93, 97f., 245–247, 250–251
Polyethylene, 2, 47, 87, 243, 245
Polyethylene naphthenate (PEN), 94, 97f.

Polyimide (film), 5–6, 15, 20, 51, 86, 94, 97f.
Post-etch punch, 7, 22, 49, 66, 68, 126, 164
Post-lamination aperaturing, 35–36
Potting, 85, 100, 238–239
    materials, 239–240
    molds, 239
Prepregs, 90, 240–205, 212, 215–216
Press pads, 133–139, 140
    hydraulic layers, 140
    release layer, 140
Press platens, 87
Process sequences, 165
    alternates:
        die stamp, 173
        PTF, 173
        sculpting, 173–174
        semi-additive, 172
    double-sided, 175
    PTH, 175
        selective plate, 182
    single-sided, 167
Progression, 35, 189, 198
    (*See also* Bookbinder)
Pumice, 108, 173, 177
Punching, 129
    tolerances, 160

Release films, 91
Repair, 16
Residual stress:
    in adhesiveless, 115
    in low-flow adhesives, 90
Resistivity, 33
    in PTF, 52
Resists; dryfilm, liquid, 144, 145
Resolution, 149
Reverse bare process, 128–169
Rigid-flex, 90, 117, 185
    bag process, 196–198
    definition, 186
    fillers, 187
    flow control, 212
    inlay/bikini, 206
    laminate voids, 188–189, 194, 214
    materials, 204
    prepregs in, 190
    sealing, 193
    sequential lam, 188
    slotting/windowing, 187
Rigid-flex design, 48

Roll reduction (for foils), 27, 104
Roll-to-roll (process), 24, 51*f.*, 88, 125–126, 138
Routing techniques, 40–41

SMA (on FPW), 40, 230
Safety factors; effect on conductor width, 34
Sales, 1
Sanders, Royden, 2
Scaling factor, 163
   offset, 163
Semiadditive process, 172
Shielding, 10, 70–73, 111
Silk-screen printing, 6, 110, 143, 245
Soda straws, 134
   sod. sulfide test for, 134
Solder wicking, 108, 135
Soldering, 225–226
   hand, 227
   heat methods for, 230
   heating methods, 230
   mass, 228
   SMA, 230
   tools/fixtures, 227–228
Standards and specs, 217
   in assembly, 225
   cost impact, 218
Steel rule dies (SRD), 7, 47, 82
   Stiffeners, 38, 58, 60, 63, 185, 221, 231
Strain relief, 38–39, 58
Stripping, 159
Subtractive process, 143
Surface finish, 8, 137, 223

Surface preparation:
   for copper, 2, 108, 134–135
   for polyimides, 203

Tear stops, 61
Teflon, 2, 15, 87, 187, 254
Tg; glass transition temperature, 90, 117
Termination technique, 38, 41
Terminations (design), 6, 29, 31, 38
Test fixtures, 161
Thermal stress, 108, 188, 190
Thermo-mechanical analysis; TMA, 209
Thermocompression, 235
Thermoplastic, 87, 89, 235, 238
Thermosetting, 20, 87, 89, 235, 238
Tolerances in design, 65
   cost consequences, 35
Tooling, 11, 45, 47, 50, 160
   in RF, 51, 160
Tooling holes, 126, 147, 250
Transparency (effect on yield), 20
Treatment; of foils, 108

Upside down process, 169

Vents (in RF), 188, 198
Vinyl, 2, 51, 86, 95, 247

Windowing (in RF), 198

Z-axis expansion, 86, 99

ABOUT THE AUTHOR

Thomas H. Stearns is president of Brander International Consultants, a firm specializing in electronic interconnection design, products, and materials. He previously developed and designed interconnections for Miraco, Inc., the Interflex Corporation, and the Lockheed-Sanders Interconnect Products Division. A noted engineer and inventor in the field of flexible wiring for more than forty years, Mr. Stearns has developed innovative rigid-flex manufacturing techniques, connectors, and controlled impedance designs.

Learning Resources
Brevard Community College
Cocoa, Florida